SISKIY(

DATE DUE

OFFICIALLY DISCARDED

THE FOREST RANGER WHO COULD

Pioneer Custodians of the
United States Forest Service
1905-1912

A Novel
by
Gilbert W. Davies

Copyright © 2003
Gilbert W. Davies

For additional copies contact:
HiStory ink Books
P.O. Box 52
Hat Creek, California 96040

All rights reserved. No part of the material protected by this copyright notice may be reproduced or utilized in any form or by any means, electronic or mechanical, including photocopying, recording, or by any informational storage and retrieval system, without written permission from the copyright owner.

ISBN: 1-887200-09-6
Library of Congress Control Number: 2003110349

Illustrations by Patricia O'Day

Printed in the United States of America by
Maverick Publications • Bend, Oregon

"The Ranger then was a doughty cuss, who chewed up nails and spit out rust"

Tom McArdle

Contents

Introduction	vii
Dedication	ix
Chapter One - 1905	1
Chapter Two - 1906	33
Chapter Three - 1907	69
Chapter Four - 1908	105
Chapter Five - 1909	141
Chapter Six - 1910	175
Chapter Seven - 1911	201
Chapter Eight - 1912	221
Epilogue	229
Notes	237
Names of People and Animals	239
Geographical Names	243
Acknowledgments	247
Bibliography	249
About the Author	251

Introduction

Although this story is a fictional account of a United States Forest Service district ranger, all the events could have happened. The problems that early rangers and the organization had with the general public were true.

The early day forest rangers were essentially custodians of the forests, protecting them against fire, poachers, exploiters and timber grazing trespassers. Some rangers patrolled an area of more than a million acres. The pine tree badge became the equivalent of a sheriff's badge. Although they stood up to their friends and neighbors who broke the rules of the agency, rarely did they use their authority in a dictatorial or tyrannical sense. The raw recruit rangers were to become foresters through the school of experience and knowledge. They came from various occupations: cowpunchers, miners, lumberjacks and merchants were included. Their chief assets were stamina, health, and a capacity for learning. They rose above mediocrity through loyalty, passion and hard work.

Rangers would have to make on-the-ground decisions continuously. Much of their territories had no phone communications so the supervisor might not hear from a ranger for several weeks. The ranger had his horse and pack animal, along with a badge, a *Use Book,* compass, map, marking hatchet, month's food supply, blank forms, notices and posters. With these, rangers had to make decisions using common sense and what was right for the local situation. They did not believe they were anything special but they believed in what they were doing and trying to accomplish.

Sometimes the job was almost too much to endure. After feeling sucked dry of blood by clouds of mosquitoes, flies and gnats, grown men were seen to sit down and weep, then grit their teeth, get up and resume whatever they were doing with added determination.

In the beginning most of them were bachelors, or if they were married, often the wife would stay someplace more civilized than the primitive forest areas of no roads, no communication, no electricity and endless loneliness.

The national forest fire guards were rugged individuals, often reared locally. For the first few years it was common practice to use a ranch or a homestead for headquarters. Usually the guard was owner of the ranch or at least lived at the place used for fire headquarters. In the early days they did not receive any formal training but had to depend entirely upon their own initiative and resources in their fire fighting activities. It was not uncommon for them to walk 15 to 20 miles to a fire and live for days on the fireline, either alone or in the company of one or two other men. They traveled by horseback or foot, carrying their tools and a small amount of supplies on which to exist for a number of days. Their cooking utensils usually consisted of a frying pan and a coffee can which was used both for making coffee and stews. Their supplies were limited to bacon, coffee, flour and beans. Sometimes they lived off the land, fishing and in some cases, killing a deer, to supply themselves with meat while on a fire.

This book relates the adventures of Buck Stonewall, a new Forest Service ranger on an undeveloped district of the Maahcooatche Forest. It tells of his work ethics, trials and tribulations. Each successful ranger had a story to tell in those days. There were dozens of them trying to slowly legitimize a new ruling government agency against a tide of public resentment for being regulated on their former free use of public land resources.

The location of the forest determined the type and amount of work each ranger did. Although Buck's district required a vast variety of tasks, even he didn't cover every feasible function. With little money and few people, the early rangers were basically custodians of the forest. They laid the firm foundation for future land managers.

Dedicated to those early forest rangers
whose indomitable spirit, passionate drive
and unwavering conviction in their mission
pioneered the way and laid the foundation
for those who followed.

1905

Recipe For A Ranger

First get a big kettle and a fire that's hot,
And when everything's ready throw into the pot,
A doctor, a miner, of lawyers a few,
At least one sheep herder and a cowboy or two.
Next add a surveyor, and right after that,
A man with horse sense, and a good diplomat.
At least one stone mason; then give it a stir,
And add to the mess one good carpenter.

A man that knows trees, and don't leave from the list
A telephone man and a fair botanist.
The next one that's added must be there, that's a cinch,
It's the man that will stay when it comes to a pinch.
Add a man that will work, and not stand round and roar,
Boil it up well and skim off the scum—
And a Ranger you'll find in the residuum.

J. B. Cammann

CHAPTER ONE

1905

"Hello, anyone home?" No answer.

"Hello there. My name is Buck Stonewall, the new forest ranger for this district. I came by to chat and introduce myself. Looks as though you have a nice, warm fire going with all that smoke coming out the chimney."

Buck was about 30 feet from the front door of the 14 by 16 cabin. He had left his 30-30 rifle with his horse tied up about 100 feet away. Forest Guard Luke Parley had warned him that Trapper Tate was not a sociable man. He had a reputation for violence, if provoked.

Luke had met Ranger Stonewall the day before at the ranger's cabin—if it could be called that, considering its dilapidated condition. The two men had hit it off right away. They had discussed the district in many aspects—personnel, geography, fire, weather and communications. After all, Luke had lived in the area all his life and Buck was the newcomer.

The next day Luke said he needed to return to his ranch and Buck asked if it would be all right if he was to tag along, see some of the district and introduce himself to Mrs. Parley. He also wanted to meet some of the other folks living in and adjacent to the reserve.

Luke said they would be very close to Trapper Tate's cabin and Buck might want to stop in. However, he also told the ranger they probably shouldn't be exposed together so he would cover Buck in case there was trouble. As they approached Tate's cabin—no one knew if Tate was his first or last name—Luke separated and vanished into the forest.

Trappers usually did not follow the profession very long. It was extremely hard work. By the time a man reached middle age he was looking to do something that required less physical exertion. They had trouble with fires as well. Cabins could be rebuilt but the fires could wipe out the production of their trapping areas. For example, the marten lived only in mature timber country and a fire could devastate its area. A trapper, on leaving the profession, would either sell his cabin and traps, move out and let someone take over or just leave.

Trapper Tate had moved to the area, built his cabin and trapped animals before the Forest Reserve Act of 1891. He was a well educated man who chose early in life to escape civilization. He was good at his profession and his pelts were of excellent quality. He did not like change and he especially did not like any government interference or restrictions. For the most part he had been left alone by federal officers but times were changing and a new administration and department had taken over the forest reserves.

"What do ya want?" came the irritated voice from inside the cabin.

"Nothing," came the reply. "I just wanted to say hello."

"Who did you say you were?"

"The new forest ranger for this district."

"Is that so! Well, Mr. Ranger I'm coming out, but it will be with a shotgun pointed right at you. So no funny stuff."

"I'm not armed and I'm not looking for any trouble."

The door slowly opened and out stepped a tall, lean, gray haired man. He looked like a figure from one of Ned Buntline's Wild West tales. Walking with a lively gait and quick movements, he held the shot gun with the barrel pointed downward.

"Mr. Ranger, what's your business?"

"Just trying to be sociable with the folks living in the national reserves." Before Trapper Tate could reply, Buck continued, "Those are some good looking traps you have hanging on your cabin. What animals are in the area?"

"Mainly beaver, marten and coons," came the reply.

"Do you have any pelts?"

"Inside and they're all for sale."

Just then Luke appeared with his horse and rifle.

It startled Tate but he settled down after realizing it was someone he knew.

"Hi Tate, Hi Buck. Thought I heard some voices down the trail. I see you two have already met."

"Well, yes and no," replied Buck. "I said who I was, but I didn't hear from our host."

"Name is Tate! Some people call me Trapper Tate but that ain't necessary. Come on in and join me in a cup of coffee or a shot of whiskey."

Chapter One - 1905

It was clear and cold outside. It felt like it would soon snow. On entering the cabin, the roaring fire and confined space made Buck and Luke feel as if they were entering a blast furnace. Buck viewed the interior curiously. In one corner was straw for a bed covered with well worn and stained blankets. In the opposite corner was a high pile of pelts. There was one chair and two wooden boxes to sit on. There was also a shelf with some books on history, music, government and animals. Quite a varied collection for this old timer, thought Buck.

"Have a seat on those boxes. Do you want coffee or whiskey?" They both requested coffee. It was black only and needed a little chewing as it went down.

"Have you read all those books, Mr. Tate?" asked Buck.

"Most of them."

"Well, if you enjoy history as much as I do, then you are an avid fan."

"Yeah, I do. Mostly about the west in the last century. Now those were the days when men were real men and women knew their place."

"What do you mean by that?" asked Luke.

"What I mean is that men worked hard to make sure their family was secure and well fed, even if it meant working twelve hours a day, seven days a week. They could go anywhere to hunt and fish, trap and travel and not worry about crime, permits, government interference or trespassing. Women would stay at home, bring up the kids, tend the garden, prepare the food and help her mate when needed. They were a team."

"Isn't that the way it is today?" commented Luke.

"No! Nowadays you see men vanishing from the responsibility of a family and the women going to work as clerks, doctors and even postmistress[1]—just like the one in Shadowcreek. Women should be either single teachers or married and stay at home."

Luke continued, "Mr. Tate, were you ever married?"

Buck knew where the conversation was headed so he quickly changed the subject and the tone.

They continued talking for awhile but the day was waning and Luke mentioned that he and Buck had a ways to go before arriving at his ranch. Farewells were said and Trapper Tate told Buck that he was welcome anytime. He also asked Buck to call him Tate instead of Mr. Tate. Although he was much older than the ranger, he felt the age difference didn't mean they had to be that formal.

Buck thanked him for his hospitality and the cup of coffee.

After riding out of earshot, Luke told Buck that he had old Tate in his rifle sights from the time the trapper opened the door to when he made himself known. Buck smiled and said nothing.

Luke asked, "How come you didn't bring up some of the new government rules?"

"Because whenever I meet someone for the first time and tell them who I am, I want to just get acquainted, be friendly and get them comfortable," replied Buck. "Now when I return to Tate's cabin with his blessings, I'll be in a much better position to tell him about my duties and responsibilities as a forest ranger and how they will affect him as a forest dweller and user."

They rode for some miles in silence as the day was losing light and the sounds of evening approached. A light snow started to fall.

About a mile from Luke's place—named Crescent Ranch—the trail widened so the riders rode abreast.

"How long have you been living at your place?" Buck asked.

"All my life. My folks settled here and worked the land. Since I was the only heir, after they both died I stayed on to live and continue to work here. I've known my wife since we went to grade school together. We both love the place. Our two sons, Matthew and Mark, are becoming a real help with all the work. In fact, two years ago, when folks at the supervisor's office asked me to become a fire guard in the forest reserve, I said no because there was too much to do at the ranch. Last year when Supervisor Kent Bolton came on the scene, he persuaded me to change my mind, since the boys were 12 and 10. Even then I said I'd work during the fire season only."

Buck looked at Luke closely. He was glad they were working together on the Maahcooatche[2] Forest Reserve. He prided himself on being able to read people and was certain Luke could be trusted and counted on in any type of situation.

Luke was a redhead with a mustache. His blue eyes sparkled and his hands were animated when he talked. At 5 feet 9 inches, he had powerful arms from years of ranch life. He was a very religious person who did not swear. His family came first, no matter what. He was a happy, contented man which is probably why he always wore a glimmer of a smile. His fire training consisted of riding into various fires in the reserve during the past four years.

By now the faint light of the Crescent Ranch loomed ahead. It was mainly a cattle and horse operation. The barn was larger than the house.

Chapter One - 1905

It was a three bedroom, white with blue trim, clapboard house. One hundred yards away was a big brown barn. A white, three rail fence surrounded the house and barn. Corrals were present along both sides of Shadowcreek Road. The dirt road ended one mile to the west. The forest reserve fire cache was in a corner of the barn. There were fruit trees, a vegetable garden and flower boxes on the window sills.

The Crescent River flowed from west to east past the ranch. Its waters originated in the high mountains of another forest reserve to the west of the Maahcooatche. It ran its course directly east before bending to the south prior to the town of Shadowcreek, completing its arc to the west and emptying into the ocean. The river flowed high and fast during the spring months. The ranch was on the north side of the river so no bridge was necessary, The steady sound of trotting horses was broken by a couple of dogs barking. The two boys and Sarah ran to meet them. Luke jumped from his sorrel horse Baron and the scene unfolded into a group hug—an unusual show of affection for the times. He turned to Buck, introduced him to his family and told them they should be extra nice to Buck since he technically was his boss. Buck told Sarah not to pay any attention to Luke's ramblings and they all had a good laugh.

Sarah Parley was a husky brown haired woman with a hearty laugh. She was an excellent cook and Buck commented that not only was it the first home cooked meal he had eaten in ages, but wondered why Luke and the boys weren't really fat. After dinner they all sat around the fire and talked. The boys wanted to hear a story so Buck regaled them with the following:

"It seems that one evening a couple of campers left their site to go fishing. When they returned they found that a bear had invaded their provisions and eaten most of their food. On further examination they found that the bear had eaten a jar of berries. It had dug a hole, placed the jar in the ground, tamped it in and then proceeded to unscrew the lid with the jar held firmly in place." This story was received with much skepticism and laughter so another was requested.

"Now this one was told to me by two witnesses," said Buck. "An old prospector left town one afternoon, according to one witness. He walked about two miles to his cabin. The second witness was already waiting in the cabin when a huge bear came between the prospector and his cabin. The bear started for the prospector. The second witness looked at his watch as the prospector turned to run back to town. His watch said 1:32 p.m. When he got to town in a panic, the first witness looked at his watch and it was 1:33 p.m." Again there was laughter and comments about embellished stories. After a couple more "suspect" stories, Mrs. Parley told the boys they should get ready for bed. Buck said he would sleep in the barn, even though his hosts objected. In an hour all was quiet at the Crescent Ranch.

* * *

Buck wanted to get an early start. He planned to visit the various ranches along the Crescent River on his way to Shadowcreek. The folks living on these ranches had property adjacent to the forest boundary and he wished to introduce himself. A road ran along the river on the north side from Crescent Ranch to town. There was also a telephone line, terminating with the Parley's. No one lived to the west of their ranch. It was a party line with the telephone exchange in Shadowcreek.

Just as they finished breakfast, the wall phone rang two longs and one short. "Hey! It's our number," said Sarah. "That's strange this time of morning." After answering hello, there was a long silence. Then she said in a calm voice, "I understand, Harvey. Tell Felicity to hang on and Luke and his boss will be there in a few minutes."

Sarah relayed the panicky message from Harvey that Felicity had slipped on the bridge and fallen into the river. She was now hung up and holding on to a snag that had wedged itself into the river's bottom about halfway between the banks. She was too far down river to be rescued from the bridge. During the relaying of the information, Buck's mind worked quickly. He told Matthew to bring him about 200 feet of three-quarter inch rope. The men ran to their horses, saddled up in record time and Matthew brought the rope. Off they galloped. Buck's mind was planning his next moves as they raced the two miles.

Harvey and Felicity Southcott lived on a ranch adjacent to the Parley's about two miles toward town. The property was connected by a narrow footbridge over a 50 foot width of the Crescent River. Their house was on the north side.

When they got to a waving Harvey, Buck took note of the situation and yelled to Luke to take one end of the rope and tie it to a cottonwood tree across the bridge, next to the bank. Since there were three trees, and two of them were down river from Mrs. Southcott, with the remaining one up river, Buck said that was the one to tie. He then tied the other end to the saddle horn on his horse Titus. He led the horse down river about 20 feet past Mrs. Southcott, making sure the rope was taut enough to pass over her head after it was tied to the tree. He removed boots, socks and hat, and put on his gloves. Next he tied a shorter rope he had with

Chapter One - 1905

him—one end to the horn and the other to his own waist. He also quickly cut the end of the long rope for a six foot section that he tied together and swung around his neck.

When all was ready he commanded Titus to stand still and told Harvey to move the horse slowly up the bank as he directed. Grabbing the rope with both hands, he swung out over the river. Even though it was drawn tight, his weight caused the rope to sag so that he hit the water with both feet. After the first shock, Buck continued to travel toward the middle of the river. When he got just below Mrs. Southcott he shouted to Harvey to move Titus up river, keeping the rope taut at all times. Slowly he approached the woman who was by this time in poor shape. She had been holding on for more than a half hour and was shivering and turning blue. Buck had figured her hands would be almost frozen so there was no way she could grasp any rope and hold on. As soon as he arrived opposite Mrs. Southcott he placed the six foot rope over her head and under her arms. In this way they couldn't be separated if she didn't have the strength to hold on during the return trip. He calmly said he was going to turn around and she should then place her arms around his neck and lock them as tightly as she could. This done, Buck slowly made his way back, reversing his movements.

Harvey and Luke grabbed Mrs. Southcott as soon as she cleared water and carried her into the house. She was shaking and softly sobbing at the same time she was thanking Buck and the Lord. Buck put on his dry socks and boots. Luke called his wife to let her know everything was fine. A large pot of hot coffee was made as the four of them sat by the kitchen table and talked about the accident. Being well wrapped in blankets, after removing her wet clothes, Felicity finally stopped shaking. She was a small, thin woman in her forties. As a hard working wife of a rancher she pulled her own weight. Under normal circumstances she would be considered quite fair. Her husband Harvey was a tall man also in his forties. He had a little stutter as he talked.

Buck and Luke took their leave; Buck to the east toward the town of Shadowcreek and Luke to the west toward Crescent Ranch. They planned to meet at Buck's so-called office in a week. It was late morning and Buck had several hours of riding to reach his destination. He relinquished any thought of stopping at ranches along the way. His pants were still wet.

Titus, a large, dark bay horse, had a steady, easy gait. Buck had acquired him almost three years before. He had finally found an animal that fit all his requirements of stamina, speed, strength and calm disposition. Titus was trained to come or go or stop by different whistle commands. He would not let another person on his back and he would stand quietly for long periods when ordered. Buck always looked after Titus's welfare before his own. His late start that morning was due to resting and cooling the bay after his two mile gallop.

*　*　*

There was not a cloud in the sky. The signs of spring were plentiful: birds busy with their nests and domestic animals with their offspring; thousands of bees and dozens of hummers taking advantage of nature's bounty of wildflowers. Buck was deep in thought. He hadn't been a ranger a week ago. The last few days seemed unreal. It had happened so fast. He wasn't even looking for a job as he passed through this part of the country. He thought about home and his life for the past 28 years. What an adventure it had been so far. He wondered if he would find the ranger's job as challenging.

Buck was born July 9, 1876, and raised on a Nebraska farm. He was christened Buck since he kicked so much in his mother's womb. He and his siblings walked miles to the one room school where they learned to read, write, speak correctly, think clearly and be tolerant of others. His favorite subject was history. His goal was college, but that time never came. An average student academically, he was a quick learner and an outstanding thinker and reasoner.

The Stonewalls were all of large frame. They were a very close-knit family and throughout the years after Buck left, he always wrote and told his folks where he was. He worshipped his older brother Ira, who left the farm at eighteen and joined the army. When word came several years later that Ira had been killed in Cuba during the Spanish-American War, the whole family was crushed. Buck never really got over it. Their sister Ivy, was eight years younger than Buck. The baby of the family was Alva.

There wasn't money for Buck to continue to college, so with the blessings of his folks, he left Nebraska for fame and fortune. Of course he knew the care and feeding of animals, how to rope, break, ride and shoe horses, use a rife and a pistol and work twelve hours a day, seven days a week without complaint. He made friends easily, gambled a little and drank a little. One time he drank too much and woke up in a strange place with no memory of the past ten hours. After that he vowed to always stay in control of his senses. He liked people, but woe to anyone who lied to him. He liked the fair sex but knew that his unsettled spirit

Chapter One - 1905

was not compatible with settling down to a domestic life. Whenever he found himself dreaming about a particular woman too much, he moved on. He had a good sense of humor and enjoyed either telling or listening to a good story.

About the only things Buck didn't like were warm beer, cold soup or fisticuffs. He thought fighting was a waste of time and energy. He never lost, however. It was probably due to the fact he didn't adhere to the Marquis of Queensberry rules and didn't apologize for it, either.

Buck's physical attributes consisted of a handsome face, dark hair, brown eyes, large hands, with no excess body fat—all this on a frame of six feet three inches and 215 pounds. His artistic assets were limited to playing the harmonica he always carried with him.

Buck headed east when he left home. He had saved some money and wanted to fulfill a dream. Being a Stonewall caused him to become an avid reader on the life of General Stonewall[3] Jackson. He traced Jackson's life from his birth in Clarksburg, Virginia, now West Virginia, to the house outside Fredericksburg, Virginia, where he had died during the war. He meditated for many minutes at the Chancellorsville Battlefield site where Jackson had been shot by his own troops.

Over the years, Buck's travels and experiences were numerous and varied. He panned for gold in Alaska; participated in rodeos throughout Texas and the Oklahoma Territory; worked his way up to ranch foreman in Colorado and Wyoming; was involved with wildfires in California; learned how to survey, build structures and hunt big game. He continued his interest in history and current events, things that helped him as a future ranger as much as the physical part.

Fifteen miles from the Southcott Ranch, Buck entered Shadowcreek, a town of 900 people.

Five miles before, the Crescent River had veered south away from the main road. The town consisted of one main road running east and west six blocks long. Side streets stopped after two blocks both ways. It was one of those "one of everything" places: one general mercantile store, one restaurant, one hotel, one community church, one weekly newspaper, one livery stable, one blacksmith, one doctor, one dentist and a post office. The town lay along Shadow Creek, a tributary of Crescent River. It was the second largest town in Queens County. The largest was Harmony, the county seat and location of the supervisor's office of the Maahcooatche Forest Reserve, 10 miles to the southeast. Commercial transportation between the two towns was by stage or train. The stage road paralleled Shadow Creek south to the village of Fish Cut at the intersection of the creek and Crescent River. It then ran east to Harmony. The railroad traveled east and then south from Shadowcreek to Harmony. Passenger trains stopped twice a day in Shadowcreek—once each way. The streets were all dirt but there were wooden sidewalks and a town park with a gazebo. The Shadowcreek Cougars School had grades from first through twelfth.

Buck knew he couldn't get to his cabin before dark so he rode directly to Ike's Livery Stable. Elroy Taylor was the owner and operator. They had met earlier. Elroy was eleven when the Civil War ended and freed his folks and his siblings. Years ago he had landed in Shadowcreek, looking for a job. His outgoing personality and strong body were what Ike Fraser was looking for, so for many years Elroy did most of the work, including blacksmithing. Ike left everything to Elroy and since there were no relatives to contest the will, it was a smooth transition.

"Hi, Mr. Stonewall," said Elroy as Buck rode up.

"How are you today, Elroy?" came the reply.

"Busy as usual, but happy to have the business."

"Could you feed and house Titus overnight?"

"Sure could. Are you ready to rent a pack animal to take to your headquarters?" Buck laughed at the reference to his headquarters.

"Yeh! Elroy, I have to buy some food, tools and supplies to take in this time. That first time a few days ago I didn't know what to expect. See you around ten in the morning. Please give Titus a good rub down."

"Sure will, Mr. Stonewall. See you at ten."

With that Buck headed for the hotel and a hot bath. Faye Wadsworth was the Pilot Hotel manager. Faye was a matronly type woman without the benefit of marriage. Somewhat short and a little overweight, she had a cheerful and caring disposition. She enjoyed some harmless teasing by her favorites.

Buck introduced himself and asked for a quiet room. It took some time to get cleaned up and ready for a bite to eat.

Mabel's Cafe was across the street. Mabel cooked and Bertha waited. Both were large and jolly ladies who talked and laughed with the customers. Buck relaxed and ate his supper slowly. Since he had stopped before on his way through, they asked what he had been doing. Buck described his office, his district and his meeting with Trapper Tate. He did not relate the Southcott incident. His narration was comical and the whole room listened and laughed. The ranger slept hard and long that night. It was a nice change sleeping in a bed instead of on the ground.

Chapter One - 1905

Next morning Buck spent an hour buying food, hand tools, fire tools, storage boxes and other necessities. On his first trip he had taken some food, blankets, compass, maps, posters and marking hatchet. This time he acquired enough food to last at least a month. When Buck phoned the supervisor's office in Harmony, Supervisor Kent Bolton wasn't in so he talked with the Deputy Supervisor Garth Kimball. He brought him up to date on his activities and his purchase of some fire tools. Although some forests did not have the funds at that time to supply every district, the Maahcooatche did have a small budget for such items. Luckily, Buck got post approval from Garth. If not, he would personally be out the money. He did have to pay for the rest of his purchases, along with some extra feed for Titus and the pack horses. Buck obtained pre-approval from the deputy supervisor to rent a couple of pack horses for three days only. He did not mention the Southcott incident. Before hanging up, he told Garth he would probably be out of touch for several weeks, but if there was an emergency then Sarah Parley should be contacted. He expected Luke to be on fire patrol.

Buck finally got his little pack train ready to go. A two mile road ran the distance from town to the border of his district. From then on it was all trail. The trails on the district were derived from old Indian, prospector, trapper and animal paths. His district was the least developed of the four districts on the Maahcooatche. There were no roads, no phone lines and no ranger station. The Parley's ranch was considered a fire guard station since it had fire tools and supplies.

Buck's office was a small, abandoned prospector's shack with a leaking roof, no flooring and no privy. Buck had been warned about the situation so he knew it would be extremely primitive living. He hadn't minded one bit. He thought back at the hundreds of times he had slept on the ground, been chilled through, gone without meals and been isolated for days. He thrived on challenges and being a forest ranger was a way to use his survival skills and creative techniques.

District 3 had it all—timber, meadows, high mountains, mining, wildlife, dozens of streams and lakes, canyons, valleys and wetlands. The reserve was divided into four districts with numbers 3 and 4 north of Crescent River and 1 and 2 south. The forest was surrounded by roads with several miles inside the boundaries of districts 1, 2 and 4. Those districts also had a ranger station office, barn and residences, plus some guard stations. District 3 was bounded on the east by Shadow Creek, a waterway large enough to be a river, on the south by ranches along the Crescent River, on the west by District 4 and on the north by private land and the Hondo Indian Reservation. Buck's district was the largest on the forest.

Before leaving town Buck thought to check for mail, in case there was some message from home or from the supervisor's office. At the west end of town he entered the small U. S. Post Office for the first time. It was a wooden structure with a simple ten foot counter running all the way to each wall. From the door to the counter took three steps. No one was present, so he thought. After waiting for a few moments he heard a woman's voice from around the corner in the back, stating she would be with him in a jiffy. The postmistress was one Mrs. Edna Lawrence. Her husband, a postmaster, had been killed in a mining accident two years earlier. She had taken over the job and was a respected and well-known figure in the community.

Buck heard the rustling of a skirt as Edna rounded the corner, carrying some mail. He was struck speechless. Riveting his gaze on the postmistress he thought, wow, what a beautiful woman. I've got to stop staring. At the same moment Edna looked straight at Buck and their eyes locked. She thought, my, what a gorgeous man. I've got to stop staring. Edna was the first to break the spell with the fluttering of eyelids and a gentle smile.

"May I help you?" she said.

Clearing his throat Buck replied. "Oh, yes ma' am, I, uh, I, uh came in to see if I had any mail."

"I don't know," exclaimed Edna. "What's your name?"

"Gosh how rude of me. Name is Buck Stonewall. I'm the new ranger on District 3 of the Maahcooatche Forest Reserve."

"You have nothing to apologize for, Mr. Stonewall. I'll check in the back." As she scurried around the corner, Buck thought again that she was a mighty fine looking female.

Edna reappeared and said, "I can't find anything for you, Mr. Stonewall."

Buck thanked her and turned to leave. Something inside him said to take it easy. This was his first meeting.

"Pardon me ma'am, may I ask your name?"

"Of course. My name is Edna Lawrence. It was very nice to meet you, Mr. Stonewall. I'll watch for any mail you receive. I guess you don't get to town very often."

"Not at this time, ma'am. I'm busy with the work on the district, but I'll be sure to stop in and check for mail every time I'm in town."

"Then I'll see you again, Mr. Stonewall."

Chapter One - 1905

"You bet, and it was sure great meeting you, ma'am."

Buck left the room in a couple of strides, hopped on Titus and headed for the forest, his thoughts deeply engrossed with the events of the past five minutes. He didn't know anything about Mrs. Lawrence, but he sure would have liked to know about her past, her likes, her feelings and her talents. He would have been amazed!

Edna was born October 15, 1881, into a large family on a ranch in Oregon. She was the eldest of eight children, the first three were girls the last four boys. As the eldest, she took on housekeeping and ranching duties at a very early age. Her mother was sickly and her father had a bad limp and was ill with emphysema. She loved the outdoors, rode horses like a pro and was a sure shot. Some of her friends called her Annie O. It was not her favorite name. Poetry was a passion. Whatever inspired her, resulted in a poem. She had stayed on the ranch until her siblings were older and her folks had passed away. They decided to sell the property and split the proceeds seven ways. One of her brothers had died as a teenager.

Edna took her money and left the area. By then she was a good looking twenty-one-year-old, five feet seven, auburn hair, blue eyes and firm body. She had graduated from both high school and Chaucer Normal School for teachers.

Before her first year teaching she had met and married Hiram Lawrence, the postmaster in the town of Ingot. They were extremely happy when their first baby was born. It was sickly and died within six months. Edna was grief stricken. She and Hiram decided to leave Ingot. He would take the next postmaster job that opened. Hence, her present status as postmistress at Shadowcreek.

By the time Buck reached the forest boundary he was reprimanding himself for his reflections about the postmistress. After all, he surmised, they had just met and he knew she hadn't given him a second thought. Little did he know!

From the boundary to his office cabin took about five and one-half hours of steady riding. Several times he would dismount and check on the loads of the two pack horses. One time he let them have a quick drink from a small stream. It was getting late when they arrived. He swiftly unpacked, hobbled the pack animals and turned them loose to forage for themselves. Titus was free to roam with no restrictions. Buck knew he wouldn't go far and with a loud whistle would come running.

Turning to the cabin, he entered and tried to light a fire. The wood was too wet. Next he brought out the recently purchased equipment needed to get rid of those damnable mice. When he had first arrived at the cabin there were dozens of mice, with no way to dispatch them except with a log or rock. It was an unsuccessful system, so he planned his future attack. From a square metal five gallon container Buck removed the top and placed several inches of water from the nearby spring in the bottom. He next strung a hefty wire through both ends of a small round can and made sure the can would rotate freely around the wire. A chunk of raw meat was skewered by one end of the wire and placed next to one end of the can. The two foot wire was placed across the top of the five gallon container with the non meat end of the can about one inch from the side. A broad stick was placed from the floor slanting up to the top of the container next to the wire. Eventually a mouse ran up the stick and stretched the inch required to get to the can which immediately rolled on the wire and dropped the scavenger into the drink. Soon there was another and another and another. After about an hour the ranger counted more than 30 drowned mice. From then on living conditions in the "mouse" house were much more tolerable.

After dinner Buck sat outside with thoughts of Edna racing through his mind. He also contemplated about his entry in the agency and the unusual way it had come about. A week ago he had been riding Titus in country to the south of Harmony. He wasn't looking for a job. Over the years he had managed to save a considerable amount of money and had deposited it faithfully in the bank near the folks' ranch in Nebraska. Since he never needed much cash, he carried very little.

It was late afternoon when he had approached Harmony. The east-west streets were identified by numbers. The north-south ones were named by letters. The train depot was on the east side of town and the tracks ran north to south. There were three churches, five saloons, a jail, county courthouse, two city parks, a six acre lake in one of them, a hospital, school, hotel, boarding house, livery stable and two mercantile stores. Most of the homes were surrounded by picket fencing.

Legend gives the origin of the town's name to the early settlers who came from all walks of life. There were miners, cattlemen and retailers. When outsiders asked how so many factions got along, the answer was they learned to live in harmony. As the center of population took root they voted to name themselves Harmony Town. The word town was later dropped.

Buck had no trouble finding the boarding house and a livery stable. Dolly at the boarding house recommended the Golden Restaurant and Saloon for dinner when Buck said he preferred eating out. After wash-

Chapter One - 1905

ing, he headed directly for the food. Sitting in a corner he faced the other customers and ordered a steak, potatoes and a cold beer. During that time he noticed two men, filthy and unkempt, who strode over and confronted a lone man eating supper. He didn't think much of it until the volume of the two men's voices rose in anger. About this time Buck's food was served and he started to eat. He rarely interfered when it wasn't his business but the loudness of the conversation made eating unpleasant. Just as he was about to go over and ask them all to please quiet down, the restaurant owner did it for him. At this time there was an explosion by the two men. The third man got up and was promptly knocked down. The owner insisted they take their problems outside. Buck was torn. On the one hand he didn't want to get involved, but on the other he didn't like the odds. Taking one last bite he got up and followed the three to the middle of the dirt road.

Buck yelled that two to one was not fair. They ignored him. He grabbed the biggest and dirtiest one and pulled him aside. Black Pete, a miner, was not going to cooperate with anything this stranger said, so he took a swing. Buck easily backed away. Then he pointed skyward and said, "Wow, just look at that." Black Pete looked up; the driving kick hit him right in the crotch; Pete leaned forward; the edge of the hand smashed him at the back of the neck; he fell flat on the ground and out. It was over in five seconds. Buck then took Pete's revolver from his holster, emptied the chambers of bullets, returned the revolver and placed the ammunition in is own pocket. During this time the lone man was getting the better of the other miner. When Lucas saw what had happened to his partner and that Buck was headed his way, he took off running as fast as his bowed legs could navigate. The fight was over. The stranger looked at Buck, thanked him and asked if he could buy him a drink. Back through the swinging doors, the two of them headed to the bar and a drink. The stranger told Buck the names of the two miners and introduced himself as Kent Bolton, the supervisor of the Maahcooatche Forest Reserve. Kent had a dimple in the middle of his chin which was the first thing people noticed. He walked fast but talked slowly.

At about 5 feet 10 inches, he weighed under 180 pounds. His disposition was pleasing and he had a well developed sense of humor. Buck introduced himself and they talked for a couple of hours. The conversation initially centered on Kent's answers to Buck's questions. Kent said the Bureau of Forestry was transferred from the Department of Interior to the Department of Agriculture on February 1, 1905. It had been a little more than three months ago and he had failed to recruit anyone for the ranger job on District 3. The other three districts already had men who had worked for Interior, like himself. They were good people and they wanted to stay even though the pay was only $75 a month. This included furnishing your own horse, feed and saddle. At this last statement Buck's eyebrows raised and he wondered aloud how anyone could make ends meet, especially if they were married with children. Kent explained that most forest reserves at that time did not have distinct districts. Some rangers were located at specific places and some of them were roving. Garth and he thought it was not a practical plan and more work could be accomplished if rangers knew they were responsible for all activities in a specific area. They had received permission to try it, so they split the forest into four distinct areas.

Buck then told the supervisor his own background. Kent said he was amazed at the amount of knowledge and experience Buck had. He asked Buck to come see him at his office the next morning. After taking his leave, Buck knew what was going to happen, so he thought about it constantly during his waking hours. He didn't know if he was ready to settle down into a full time job. The next morning Buck stood in front of the leased dark red brick one story building on the corner of 5th and E streets with a sign out front stating Maahcooatche National Reserve Supervisor's Office. It was a three room building with the reception area split from the clerical area by a long counter and a low swinging door at one end. The supervisor and deputy each had a room. Out back was a separate storage warehouse type structure. The interior was all wood of various species. Wooden files, desks, chairs, shelves and waste baskets were organized and convenient. A public corridor led to the other two rooms.

The first person Buck saw was Lucy Neville, the office clerk. Lucy was in her mid-forties and the model of efficiency. When both supervisors were gone, she ran the office and answered questions for the public. She wore glasses and her dark hair was tied into a bun. With graceful movements and a lovely smile, it was apparent she was a loyal and caring employee. Buck and Lucy introduced themselves and Buck was ushered into Kent's office. Kent introduced Buck to Deputy Supervisor Garth Kimball and the three of them talked seriously. Garth was a heavily built man about six feet. With light hair and bull neck he was built like a heavyweight boxer. With all this went large hands and feet. Buck wanted to know, if he accepted the position, what was expected of him—since he had had no previous experience as a ranger. They assured him that with his background and obvious drive, he would succeed. They ex-

Chapter One - 1905

plained the various functions of the Bureau of Forestry. They also said his main job during 1905 would be to patrol for fires, watch for illegal timber removal and poachers, talk to the folks living inside and outside the reserve boundaries and locate a site for a permanent ranger station to be built after July 1st, if the funds were approved. They showed him a map of the geographical features on the district and gave him a crash course on various policies, regulations and statutes. He read pages of official directives, letters and claims. At the end of the day Buck agreed to try the job for one year. He said that if at the end of that time either party was unhappy with the other, they should separate on amiable terms. Everyone was happy with the agreement.

Buck did ask how they could make him a ranger without going through the civil service. They responded that there had been no takers for the position so they obtained special permission to hire someone locally. At that time Buck was local. They also explained that he wouldn't have to come to Harmony often since he could send and receive mail, supplies and phone calls from Shadowcreek. Buck thought that situation was a big plus. They told him about the fire guard Luke, and that he was a good man. Finally, he filled out some personnel paperwork, received a badge and exited with comments regarding lots of luck.

Buck sat, still thinking about the past few days. The three horses had not strayed and as darkness came he got up, stretched and headed for a corner of the cabin with a blanket for a cover.

* * *

For the next few weeks Buck and Luke trekked all over the district. There were no fires yet. They found what they believed to be a perfect spot for a new ranger station. It was about equal distance from the Crescent Ranch and Shadowcreek. In fact, when the three were tied together on a map it almost made a perfect triangle. It was at a lower elevation from the current cabin; it had plenty of natural feed for animals, plenty of convenient water and was flat enough to build a barn and large corral, plus the office. A spring and a couple of good sized lakes and several swift mountain streams were nearby. Its general location on the district map showed the word Lakefield. As far as Buck and Luke were concerned, they unofficially called their district Lakefield. District 3 sounded too impersonal.

Luke told Buck there would be a big July 4th celebration in Shadowcreek. Since he hadn't been to town in several weeks, the ranger told his guard to keep on fire patrol and he would check out what had been going on in the supervisor's office and get any new instructions.

Buck arrived in town July 3rd and called the office before they closed for the holiday. Lucy said she was hoping to hear from him and that he was to report first thing on the morning of July 5th to the supervisor. She didn't know what it was about.

Edna was at the fireworks display and Buck sidled up to her. She was very glad to see him and asked how he had been and if he was in good health and happy with his new job. Buck had almost forgotten how beautiful and charming she was. He kicked himself for staying away so long, although he knew that his work came first. Buck told her he had to leave for Harmony but would be coming back through in a few days. They continued to talk for another hour.

Buck was at the office when Lucy opened up. They made small talk until Kent showed up and ushered Buck to his room. Kent listened to a thorough report from District 3. Buck said that he had become acquainted with the district geographically and with its resources. The most exciting event was that Luke and he had found what they believed to be a perfect spot for a ranger station. Kent listened intently without comment. Finally Buck mentioned that he referred to the district as Lakefield and he surely hoped the supervisor didn't mind. Kent smiled and said it was a good name.

"That's about it," said Buck. "I'm really curious as to why you wanted me in person this morning."

"I have some great news," answered Kent, "but before continuing I'd like to wait for the other three rangers to arrive. I asked you to come early because you've been out of touch and I wanted to hear a blow by blow account of your activities. They should be here any moment." Kent looked at his watch and before he could put it back there was the sound of a door opening.

"Come on in you two," said Kent, "I want you to meet your new neighbor." The two men stood and the supervisor introduced the ranger on District 1, Nate Bennett and District 2, Ralph Hempstead, to Buck Stonewall, District 3 ranger. He added, "Or should I say the new ranger on the Lakefield District?" Nate was short and Ralph was tall. They both were friendly. Buck sensed there was a feeling of unhappiness with Nate.

Hugh Tanner, the District 4 ranger, had not arrived. His district headquarters was farther from Harmony than were the other three. Although he could travel all the way by road, there were often unexpected delays. In

Chapter One - 1905

about ten minutes there was a loud laugh from Lucy and a louder one from Hugh Tanner as he entered the front door. He was a jovial character who seemed to accept life as it came. He was a well-proportioned six footer with a muscular body and long mustache. After more introductions, the supervisor said the meeting should come to order.

"Gentlemen, I have some exciting news." Kent was smiling. "On July 1st everything changed. We are no longer the Bureau of Forestry. We are now officially the United States Forest Service. We will remain in the Department of Agriculture and our forester[4] will continue to be Mr. Gifford Pinchot. As Mr. Pinchot has said in the past, he insists that we remain a decentralized organization. When we were in the Interior, if you remember, we had a manual to refer to. Now, before me, I have copies of our new *Use Book* for each ranger's use. It is designed to fit into your shirt pocket and you are to keep it with you at all times. Any questions?" There were several questions as the supervisor handed out the 142 page, four and one-quarter by six and three-quarter inch book. The four rangers thumbed through it and there was a general discussion that lasted until noon.

The six of them—Garth Kimball had joined the group—proceeded to have an animated meal. Now each ranger could slip everything he needed to know in one pocket and patrol his district on horseback.

After lunch the supervisor asked Buck to return to his office. He said there was some more good news that affected the Lakefield District.

At the office Kent continued. "I received word that we have been appropriated $445.15 to build a ranger station on your district. You will have to build it but I will help at this end if needed. You will need to keep track of all expenses and not exceed the amount. It will be your responsibility to get the office built and if there is any money left over, then do the corral and barn. For now your sleeping area will be in the office. One more thing, you will have to spend all the money before June 30, 1906. This is because it comes from an annual appropriation. Well, Buck, what do you think? Can you do it?"

Buck gave a typical reply, "Wow! That's good news. It's a tall order but I'm sure, with Luke's help, we can do it. I'll start working on the plans right away, figure the needs and head back to Lakefield. I'd like to stay here, use the office and get your approval before any money is spent."

"That's great. I figured you would welcome the news."

"One more thing, Kent. Unless it's not available, I'd like to obtain all the material and supplies at Shadowcreek instead of Harmony. It's much better logistically."

"Sure thing, Buck. I'll leave you alone now. And thanks a lot for your willingness."

For the rest of that day Buck worked on drawing the plans and layout of the ranger station. Measurements and amounts down to the last detail were worked out. He wanted to make it a showcase for the new U. S. Forest Service Lakefield District.

Buck contacted Elroy in Shadowcreek and reserved a string of horses. (At that time Elroy didn't have mules). He did the same with the mercantile store on materials. They said if they didn't have it they would order it. After receiving the boss's approval, he made his farewells, retrieved Titus from the local livery stable and headed toward Shadowcreek with numerous thoughts spinning in his head.

The previous evening in his room at the boarding house, Buck had read the 142 pages of the *Use Book*. He now had a tool to refer to when talking with the public. It was written in a positive tone instead of always prohibiting something. He recognized how intelligent that was.

At Shadowcreek, Buck brought Edna up to date. He asked her to join him for dinner at Mabel's Cafe and she accepted. They had a pleasant meal but Buck knew he had a long day ahead of him and so he departed after escorting Edna to her front door. Edna's house was a one story, light blue with cream trim structure. It had some gingerbread featured on the outside with a small garden in the rear. The front was surrounded by a yellow picket fence and the back was enclosed by 1 by 12 inch boards. Flowers were in window boxes and along the walkway to the front porch.

Buck didn't leave for his district until past noon the next day. He had horses, packing, purchasing food, materials, supplies and a myriad of other things to take care of. As he left town leading the pack, Edna came out of the post office, waved and wished him God speed.

Before leaving, Buck called the Crescent Ranch. Luckily Luke was home. Buck gave a quick rundown and asked Luke to meet him at the Lakefield site. If Luke got there first, he was to locate some trees where they could make a small corral to hold all the animals. If Luke could, he was to cut down some saplings and use them for a one rail fence.

As the pack train slowly headed into the forest, Buck continued to stop and readjust the loads. It was slow going. Luke had been there for an hour and a half when Buck arrived at the site. They immediately unloaded the animals, led them to water and left them to forage in the corral that Luke had hastily erected. Buck was very appreciative of Luke's work. They identified the exact spot for the combination office and resi-

Chapter One - 1905

dence. It was to be a one room, four windowed structure, 16 by 20 with a loft and a porch. Buck had purchased a tool sharpener, measure, adze, level, ax, mattock, hammer, one and two man crosscut saws, sledge hammer, wedges and all the other necessary items he could think of. He moved out of his old cabin to live in a tent until the station was built.

* * *

During the next two months, dozens of logs were cut, peeled and adzed by Buck—with Luke's help when he wasn't checking for fire from Tip Top Mountain or Florin Peak. They figured that the lack of fires to date was the result of an above normal snow pack during the 1904-5 winter. Lightning was their biggest worry.

They found trees close enough so that Titus and Baron could drag them to the Lakefield site. Buck decided to peel all the logs, resulting in fewer bugs and less moisture. Each log was rough hewn, since only a small amount of material would be removed from both sides, plus it could be done after the log was in place. The corners were made with a single middle notch on the bottom. This left the ends of the logs projected beyond the corner. The single type also drained the water better than either the double or top saddle notch style. The roof was ridgepole construction. Buck hadn't made up his mind about what chinking to use to fill the spaces between logs. He also would wait to build a wooden floor if there was enough money left. The barn, though slightly smaller, was to be built with similar material. Buck didn't always use living tree logs. There were hundreds of fire-killed snags across the area. Those that were still solid were used on the barn and for wood in cooking and heating fires.

As the work progressed Buck had to be inventive. Without the luxury of working with unlimited resources and tools close to town, he jury-rigged problem situations that continued to arise. His past experiences were a godsend.

Often, after working a full ten hours, Buck would tramp up the hill to Green Lake and catch a couple of rainbow trout for dinner. This was his way to relax and think.

As the walls went up, it was necessary for Luke to be present. Using the broad ax, adze and saws, the work was slow and deliberate. Neither man carried a watch. They weren't concerned with time. They wanted to get the station completed before the snow came but they also wanted to do a quality job.

Buck's plan included a corral, barn and outhouse, in that order. He decided if there wasn't enough money to do it all, he would dip into his own funds. Except for fire patrol, the two men put aside other work. They were determined to get the shake roof on before winter. There were enough cedar trees in the area so the ranger decided to make the shakes himself.

One day Buck and Luke were busy lifting logs into place on a wall. Engrossed in their work, they didn't notice a black bear ambling up the trail. A neigh from Titus alerted them to its presence. Luke hollered and threw a rock. The bear took off but came back. Buck grabbed his rifle and shot over the top of the bear's head. It didn't run off. The men got on top of the highest wall and waited. Twenty minutes later the bear left.

Since Lakefield was so isolated, it had few visitors. At about noon, a week after the bear incident, Deputy Supervisor Garth Kimball rode in with Hugh Tanner, the ranger from District 4. Their destination was Hugh's station at a place named Barnesville. They decided to check on the progress of the cabin. Buck and Luke were working on the roof and it was hard going with just the two of them.

After lunch Buck reminded Garth that his boss had said he would help if he could. Buck figured that since Garth represented the supervisor, it was the perfect time to collect on his promise. Garth and Hugh were not dressed for the occasion but the four of them got the roof up and covered in one day. On leaving to continue their trip, Garth told Buck that if he treated all his guests that way, he wouldn't have many visitors. Buck said that was the way he liked it and that when strangers came by in the future he would put them to work on all sorts of projects. This jesting went on for a few minutes and the two visitors departed down the trail in a jolly mood.

The next day the porch went up. The only things missing on the cabin were the door, windows and floor. The loft had been framed but also lacked a floor. Buck figured about two-thirds of the funds were left. He prioritized the list of material needed to complete the cabin, build the corral and barn and finally, a privy. Figured in was the cost of the pack train for at least four trips, since there would be lumber, doors and windows. By this time he decided to use wooden sealing boards for chinking. The unwieldy lumber would mean both men would be involved with the pack animals—one in front and one at the rear. Using his plans, he had ordered the doors, windows, hardware and other necessities from the mercantile store in Shadowcreek at the beginning of the project.

Chapter One - 1905

The first trips went smoothly. Doors, windows and some lumber were brought in. The fourth and last trip was almost all lumber for floors and fence. Then it happened! Within 100 yards of the station site, the pack horse directly behind Titus stepped on a yellow jacket nest. Chaos ensued for the next few seconds. The horse jumped, bucked, thrashed and whinnied. The inevitable result was that the lumber load broke from the animal and crashed on the trail. The yellow jackets continued their assault. The other pack horses whirled and broke loose with their loads coming off as they wildly dashed about. Titus and Baron remained calm but Buck and Luke had their hands full trying to corral the other wild-eyed equines. It took a full two hours to get things in order. A couple of the horses took off toward town and had to be "captured." Another one lit out across country to the north. When all was in order, the two men sat down exhausted. There was lumber and material strewn over three acres. Several pieces of lumber were broken. It took them the rest of that day and some of the next to gather and inventory all the cargo.

They fished that afternoon and the delicious taste of fresh trout made the previous day and a half easier to accept. Luke would return the horses the next day and no more pack trains would be hired that year.

The installation of doors, windows and both floors to the cabin was completed in two days. Buck had previously brought in and piled enough logs to make a barn. However, he wanted to get the corral up first and this took another day and a half. With the two of them working, the barn was finished in two weeks. They then nailed the horizontal strips of wood over the chinks in the cabin and barn.

The funds were depleted. There wasn't enough left for the outhouse. Since Buck would be the main user of such an edifice, he reasoned that he should pay for it personally. He also didn't want to go through a winter without one.

After putting on the finishing touches the two men congratulated themselves on a job well done and decided to take the next day off. It rained for the first time that fall. They had brought some dry wood into the cabin earlier and with a roaring fire and enclosed building they were dry, warm and comfortable.

* * *

The thunderstorm had brought lightning. They had to do some serious patrolling. Tip Top Mountain was the highest point on the district. The timberline was below the summit so they had a 360 degree unob-

Chapter One - 1905

structed view of the country. Luke said it was a sacred mountain to the Hondo Indians and they held gatherings at the summit at regular yearly intervals. However, he had never been in the vicinity when these affairs took place.

There they were—three smokes in the area of Pondosa Ridge. They agreed to split up. Buck took the fire to the north, Luke the one to the south and both would converge to the middle one when their own was out. Both men carried an ax, shovel and saw. Buck had some food, canteen, gloves and one blanket. The food consisted mainly of a slab of bacon and beans. It took two hours to find "his" fire. It was a single large fir snag. He made a wide fireline around the base and burned the area out. No sooner had he sat down to rest when the snag crashed, skidding down slope, scattering fire for several hundred feet. Buck jumped on it. He started firelines on both sides of the burning debris and managed to control the slop-over just before dark. With an eye open for problems, he slept fitfully until dawn. All the next day he threw dirt and crawled on his hands and knees probing for hot spots over the whole area. One more night at the site and Buck was satisfied that "his" fire was dead out.

He headed south to where his memory took him to the location of the middle fire. Luke was already there. His first fire had been a punky log. No trees were in the vicinity. It was smoking in an open area in the middle of a large meadow. He spent an hour on containment and more than a day to make sure it was completely out.

Both men analyzed the present fire. A lightning strike had hit a sugar pine and caused a left spiral split of the smoking trunk descending from the top to the ground. They considered several alternatives but decided to fell the tree. It was a dangerous maneuver. Ax vibrations could send down burning bark. The tree might disintegrate. If they sawed it they weren't sure which way it would fall. The land was flat, so if the tree did come down, it wouldn't go anywhere. Buck took his ax and asked Luke to stand away and warn him of any falling material. The ax hit the tree, Luke shouted and Buck ran. It had been burning inside for more than three days and when the blow hit the outside, the tree collapsed in a heap.

They both worked furiously to contain the scattered embers with about six hours of daylight left. The next day, after feeling for hot spots until the middle of the afternoon, they left. Luke headed for home and Buck to the new station. Quite a pile of wood for heat and cooking for the winter months were needed and Buck knew that other district problems had surfaced during his full-time construction work. There were some trespass cases to look at, plus several of grazing and mining inquiries. Boundary surveying was also on the agenda. He thought he could catch up during the winter.

Buck wondered how the supervisor's office would rank the work. However, he surmised they had hired him to be the ranger and expected him to be on his own and make his own decisions.

Buck realized he needed to go to Harmony and report on the three fires. Both he and Luke considered themselves fortunate for controlling all the fires in such a few days. He also needed to report on the completion of the Lakefield Ranger Station. He would tell Kent that he spent every last cent and finished the privy with his own funds. He thought about making a joke of the fact that the government was too cheap to spend money on human needs.

As always Buck headed for the post office. Edna was extremely glad to see him. She asked for a complete rundown of his activities for the past few weeks. She listened intently about the "accident" and was engrossed by the fire adventures.

Edna told Buck she thought he was an amazing man and the Forest Service was lucky to have him. He blushed but the color was mostly hidden by his stubble. He hadn't shaved in days. Mumbling some response he asked Edna if she would dine with him that evening. They had agreed to call each other by first names. After a delicious dinner at Mabel's Cafe, they talked and talked. It was late when Buck fell into bed at the Pilot Hotel.

All of Buck's plans for his winter months of work changed the next day. He phoned the supervisor's office that morning and Lucy said he was to come immediately to the office by Kent's orders. With Titus doing the walking, Buck had time to think about what Lucy said. His thoughts wandered from: he was going to get fired; he was going to get chewed out for neglecting some work on building activities; he should have been in contact more frequently; he was out of line for asking the assistant forest supervisor and district ranger to help him; he was going to be transferred. In the end he justified his wild thoughts by thinking if they didn't believe he was up to the job he would just move on. After all, he had said he would take it for only a year and then decide. They had agreed to the proposal but if they wanted to move things up by several months it was OK with him. He hoped the new ranger would enjoy "his" outhouse.

Chapter One - 1905

By the time Buck stationed Titus at the front of the office, he had settled down. Lucy waved him into Kent's office. The supervisor's greeting was reserved though he seemed pleased that Buck was finally there. Kent got right to business.

"Garth tells me you are doing an outstanding job of building the Lakefield Station office." Kent went on, "I thank you for your work. However, winter is fast approaching and much work that should have been looked into wasn't. I realize you are new to the Forest Service and have a lot to learn. Mr. Pinchot is more concerned with having men who are experienced in field work instead of book work. His office sent down word that no outside work would be permitted and rangers need to deal tactfully with all classes of people."

Buck started to say something but Kent went on, "Buck, I'm not concerned with your lack of experience or your report writing. You have had more experience in a variety of things than all three of my other rangers combined. I've been placed in a difficult position by my superiors. About three months after your employment paperwork was submitted, I received a call from the head of personnel at Washington. He said that in 1906 there would be some ranger exams given for new rangers. They would have to pass these exams before they could be hired. He specifically referred to you, Buck. I reminded him that I had a waiver for District 3, due to the difficulty of filling the position. He countered by stating that the waiver was for 1905 only and that in 1906 the ranger on my District 3 would have to pass the exam. I argued some more but he was adamant." Buck listened intently without expression.

Kent went on. "I want to apologize to you for this. I consider myself a man of my word and I know what I said to you about the waiver. I am sorry. Please, before you say anything, understand that I really want you to stay on as ranger at Lakefield. I have no doubt that you can pass the exam next year. I'll accept your decision. So what do you think?"

Buck didn't answer for a few seconds. Then he said, "Kent, I know that you're sincere in your feelings and I accept your apology. It wasn't your fault. As for the exam, I don't have a problem with it. I believe I can pass it. What I am most concerned with is my work in the field. It has been neglected and I feel responsible. I promised you I would stay at least one year and then decide about staying on permanently. The year won't be up until next spring, so I'll stay at least until then. I know I would like the field work in grazing, timber, mining, trails, communications and all the rest. I also have a personal reason for staying on the Lakefield District. You probably can guess what that means. I like the people I've met and I like the challenge of helping to develop the reserve."

A look of relief crossed Kent's face as he listened. He was smiling as Buck finished.

Kent responded, "Thanks so much for your attitude, Buck. Now I have something to tell you that will make you really happy. We've got the money to pay for your horse feed when your cost exceeds $75 a year so be sure to submit your paid bills to Lucy whenever you get to town. I've been thinking about your field work as you have. But I've also been thinking about something else. I didn't tell you right away because I didn't know what you'd say, though I admit you said what I expected. Anyway, the Maahcooatche has grown in size and your district has the largest increase. President Roosevelt signed a proclamation listing additional acreages and boundaries to the reserve. This means that you are now further behind in work than before and that you have more patrolling and responsibility."

"Where's the added acreage?" Buck inquired.

"On your district, it's entirely toward the east. From the current boundary of Shadow Creek it extends eastward for several miles. The TL& W Railroad tracks split the area from south to north. There is some checkerboard ownership involved so it will be more difficult than if the Forest Service administered all the land. The railroad sections of land will be protected and included as part of our fire suppression activities. Another possible additional workload you may have is fire trespass situations. Trains throw off sparks and embers and get hotboxes. When this happens we must keep track of all fire suppression costs and bill the railroad. However, your biggest concern will be from grazing problems. There are several ranches adjacent to the new boundary and they have for many years used the area within the new reserve for forage without a permit. You'll have to be on your toes when you talk to the ranchers."

"Well, I've talked to ranchers all my life," said Buck, "and I usually get along fine with them."

"I'm sure you will, Buck. I have plenty of confidence in your skills. Here's a copy of your new district boundaries. You now have more than 300,000 acres to patrol. For your information, there were changes in District 2 and 4 also. Again congratulations for building the Lakefield Station. I hope to visit there before winter sets in."

"Thanks! I spent all the money and submitted the paperwork to Lucy, along with three fire reports. Now don't get me wrong on this, but I did finance the lumber and hardware for the latrine out of my own funds, but I'm not asking to be reimbursed."

Kent laughed and said, "Yeh! The Forest Service is on a tight string. If I can find any money by the end of the fiscal year then you'll get some or all of your contribution back."

"Well, I'm not expecting it," answered Buck.

They both talked for another half-hour about district work and potential problems. Kent reminded Buck to stay in touch as often as possible. He also said that Luke would be off the payroll next month and wouldn't return until spring. However, the phone at Crescent Ranch could still be used. His last comment was that the ranger was to make his own decisions on the ground and that he would stand behind whatever was decided. Buck thanked him for this as they shook hands. Buck said goodbye to Garth and Lucy and departed, then headed straight for Shadowcreek and a bite to eat.

The new acreage was at a lower elevation than the west side of the district. Buck spent the next couple of weeks discovering his new domain. He stopped at the cattle ranches along the way, introduced himself, but didn't get involved with permits and all the other requirements that eventually would be needed with grazing private stock on government land. He planned to come back and do all of that.

* * *

It was December and the pile of wood for the fireplace was being depleted at an alarming rate. Buck spent two days splitting and piling dead and down trees.

He hadn't seen Luke and his family for some time, so instead of riding to town he headed for Crescent Ranch. When he arrived only Sarah was there. She was glad to see him and said that Luke was in town on some business. He would be picking the children up on his way back. She said that Edna had called a few days before and said he had some mail. Buck thanked her and rang the post office.

"Hello, Edna, this is Buck. How are you doing? I've missed you. I'm calling from Luke's and Sarah's place. Sarah said you had some mail for me?"

"Sure glad you called Buck. I've missed you too. Yes, I've got a first class letter from Nebraska. It says urgent on the outside. When will you be picking it up?"

"Oh no!" Buck exclaimed, "Edna, if it's not illegal would you please open it up and read it to me?"

"I will if you don't mind all the other folks on your party line knowing about your personal affairs."

"That's OK, I've nothing to hide. I'm worried that something has happened at home." Edna read the letter slowly. It was from Ivy. She wrote that the folks were very sick and didn't know if they would live through the winter. She asked for Buck to come right home, if possible. She had recently married and her husband, Henry Springer, along with Alva, were still at the farm. As soon as Buck knew when he would arrive, to let her know by telephone.

When Edna was through, Buck told her he was on his way to town and would she please get the information about when the train would come through and what transfers it would take. He then hung up, quickly told Sarah what had happened and was on his way.

At the outskirts of town, Buck met Luke and told him the news. He asked Luke to call Kent and tell him what had happened and that he would call him upon his arrival in Nebraska. Then he went to the post office.

Edna gave him the information. The train would arrive in twenty minutes. Buck hugged her and said he would have to hurry to Ike's Livery to leave Titus. He promised he would write.

Elroy was at the stable and said he would take good care of Titus. Buck wasn't to worry. Buck thanked him and headed for the train depot. He bought a ticket all the way to his home town in Nebraska. The train was on time. Buck found a seat and was deep in thought as it left the station.

1906

The Prodigal

I was tired of the silence and grandeur,
Of the solemn, unchanging hills,
Where the only echo of music
Was the splashing of mountain rills
I heard in my dreams in the cabin,
Lonely, and lonesome, alone,
The hum of the far-away cities
Insistently calling me home.

Now I dream in a twenty-room building
Of the men and the days back there;
The work that was always man's work—
The tang of the mountain air.
These are pretty good fellows
As men in the cities go;
But those clear-eyed, weather-bronzed rangers
Are the sort I'd rather know.

So I think I'll go back to the Service
I'm sick of this routine work.
The monotony's driving me loco;
I wasn't cut out for a clerk.
Out there where the Rangers are waiting;
Out there in the limitless open;
There's a job that is more to my style.

Jack Welch

CHAPTER TWO

1906

Buck wept unashamedly. He was alone as the barn animals watched curiously. His stoic behavior since arriving at the Nebraska farm and during the funeral services of both parents had ended. The news that greeted him when he had first arrived was that his dad had died the day before. His siblings were completely broken up and counted on Buck as the older brother to take control. His mom was not expected to live much longer. Both had influenza.

Friends and neighbors called, brought food, worked around the farm and asked if they could help in anyway. The Stonewalls were appreciative and thankful. The night their mother died, the three of them stayed in her room. The doctor had said there was nothing more he could do. The next morning Buck made arrangements for her funeral just as he had done for his father's. He kept telling himself he needed to keep calm and think clearly. He was the rock that Ivy and Alva needed to lean on.

When the second funeral was over, Buck excused himself and headed for the barn. He not only wanted to be alone, he wanted to pray and meditate. Regaining a semblance of exposure and wiping his eyes, Buck stayed in the barn until nightfall.

Memories of his parents and siblings and growing up at the farm and school came rushing back. He remembered his dad taking him fishing and hunting; his mother cooking and keeping his clothes mended and clean; his older brother Ira helping him in so many ways; the family sitting down to supper and being thankful for what they had; his teachers pushing him into being a clear thinking student; and all the farm animals and wild critters teaching him a reverence for life. He made peace with himself and God.

Now it was up to him to help his brother and sister decide their future. His folks' will specified that all assets were to be divided as equally as possible among Buck, Ivy and Alva. Buck would never demand his third. To do so would have required selling the farm.

A month later, Buck was on the train returning to Shadowcreek. It was resolved that Ivy and her new husband, along with Alva, would live and work on the farm. When there was enough money to help pay for Buck's partial interest, they were to send it to his bank. In the meantime Buck transferred his accounts to the bank in Harmony. He made no demands of time or amount of interest on his siblings. The farm was theirs and he was not going to interfere.

As the train rolled westward through farms and mountains, Buck thought about all that had taken place since boarding the train in December. It was now February. He had written Edna several letters. He had talked to his boss and was relieved that Kent had not only heard from Luke that first day but had told him to take all the time he needed. It was winter and nothing was that urgent. He thought about what terrific folks were in the Forest Service and how they all seemed to work toward the same goal and try to assist and be good neighbors and reasonable with the public.

It was dark when Buck arrived in Shadowcreek. Although the days were getting longer, they were still short. He was famished but food could wait. The livery stable was closed so he walked to Edna's house. His heart was pounding as he knocked. The door opened slowly and there she stood. Neither said a word. They just melted into each other's arms and held on tightly. They both spoke at the same time and then laughed and hugged again.

Edna wanted to hear all that had happened to Buck, and Buck wanted the same from Edna. They took turns. It was an hour later when Edna said, "Buck, I bet you're hungry." Buck hadn't thought about his empty stomach since he knocked on the door. Edna had eaten earlier but insisted that Buck stay and eat some home cooking. After his fill, they talked some more until it was time for Buck to obtain a room at the hotel. Faye Wadsworth had a strict rule that no rooms were rented after midnight. After more embraces, Buck took his leave, procured a room and fell into dreamland, with Edna in his thoughts.

It was time to report for work. Titus pranced and whinnied at Buck's arrival. As he trotted down the road to Fish Cut on the way to Harmony, he kept looking back to make sure Buck was still on top. Buck thanked Elroy Taylor for taking good care of Titus and paid him a handsome amount. Elroy was elated and grateful.

During his absence Buck decided to obtain a pack mule. He had thought about the difference it would have made during construction of the Lakefield Station. It would be one of the first things he did on his return. Since the station now had a barn and corral, there was little reason not to have one. Although Elroy did not yet rent mules, he told Buck about one that was for sale. He knew the owner and knew the animal was well bred, well fed and well treated. Buck stopped at a ranch on the outskirts of Fish Cut. He introduced himself and inquired of the mule. Yes, it was for sale and was a fine specimen. The two made a deal and Buck said he would pick it upon his return from Harmony. He gave it the name Stub, figuring it was automatically stubborn like all mules.

Lucy gave Buck a warm welcome. She said Garth had gone to help Nate Bennett, the District 1 ranger, regarding a timber trespass. Kent was due in any moment so it was all right for Buck to wait in his office. As he waited, he looked through some files marked District 3 on the desk. He had cleared it with Lucy. They contained information about a railroad fire, the report of a saloon on government land and a phone line. The fire had been controlled and a billing to the railroad was made; the phone line was a request for funds to string a line from Lakefield Station to either the Crescent Ranch Guard Station or the exchange in Shadowcreek; the saloon problem was marked with a notation that the ranger would attend to it as soon as he returned from Nebraska.

Another folder was marked Ranger Examination. Again with Lucy's permission he looked at it. His name was listed to take the exam in April in Harmony along with 15 others. The only name he recognized was Nate Bennett. Although Nate had been employed by the Department of Interior, he was having a difficult time adjusting to the Forest Service system in the Department of Agriculture. The policies and directives that filtered down from Gifford Pinchot were not for everyone. Nate had a second job that he was forced to relinquish. Rangers were also being delegated more administrative work. The public was to turn to the ranger instead of the forest supervisor for answers to problems. It was all evidently a bit much for Nate and he was required to pass the exam in order to continue work with the Forest Service.

Kent entered and greeted Buck with a warm handshake. He asked him about his trip and if he wanted to continue as ranger for the Lakefield District. With Buck's affirmative answer, plus his statement he would like to make a career of the Forest Service provided he passed the exam, Kent was delighted. The supervisor outlined what had happened on Buck's district during the past two months. They both agreed that better communications were needed. Buck was to assess the best route for a phone line from the Lakefield Station to the nearest and cheapest access point.

So there would be no misunderstanding, Kent ordered Buck to close down a saloon built on Forest Service land in the northeast corner of the district. Buck's only question was to ask if he should issue the owner a citation or just have him remove the saloon. Buck was to use his own judgment.

They talked about the upcoming ranger examination. It would take two days, according to Kent. Buck was to report to the supervisor's office the day before it began.

As Buck was about to leave, Kent reminded him to make sure all the ranchers who grazed their animals on Forest Service land were aware of new fees and required permits. They shook hands and Buck departed. He thought about the logistics of performing the work. Since the saloon location and the ranchers were in the same general area, he decided to pick up Stub, pack him with food and supplies and head north. After that he would swing back and open up the Lakefield Station. He was a bit concerned since the station had not been properly winterized, due to his quick departure from Crescent Ranch in December.

Stub performed well. Buck purchased all the necessary items for a diamond hitch and they were finally off with food supplies for two weeks. It was surely much easier than loading food and supplies on Elroy's pack horses and Titus. Buck wondered why he hadn't used a pack mule from the start.

A breeze refreshed them. The pines, firs, cedars and hardwoods rustled and swayed. A few clouds were outlined against the blue sky. Birds sang and rodents scattered as the little group continued steadily along a primitive trail. Buck could have taken the longer route along the main road north from town but he chose to stay in the forest on the west side of Shadow Creek. He knew there were no bridges in the places he would need to cross. When he became acquainted with the addition to the district the year before, he had used some low water crossings. However, the water level was higher in the winter and spring.

They reached Morgan's Meadow as the sun was descending to the earth's horizon. There was about an hour left of decent light. Buck figured that Stub would stay close because there was plenty of feed and water at the meadow and he was hobbled. Although snow was present there were places where Buck could pick up dry kindling and chop wood fuel. He soon had a crackling fire and heated some beans and bacon for supper in his small spider.[5] It was cold but he had brought two blankets instead of the usual one. The night brought sounds of crickets chirping and frogs croaking.

About an inch of snow had fallen. Buck didn't give it much thought

Chapter Two - 1906

as he prepared to load up for a busy day. Stub hadn't gone far and Titus was there a few moments after Buck's whistle. Before leaving the area he made sure the campfire was dead out. With a last look at the meadow, he thought it would be a perfect location for a future fire guard station or campground.

As the trail led north it also veered east toward Shadow Creek. The water was too high to attempt a crossing. Buck continued northward for the next couple of hours. He entered an area of his district that was unfamiliar. Soon he imagined would come the forest boundary. He knew there was a road that branched off to the west of the main route from Shadowcreek, so he figured he could backtrack over the bridge to the main road on the east side of the creek.

Suddenly he heard a strange sound but couldn't pin point what or where it was. It seemed to become louder as he rode. He stopped abruptly and stared in disbelief. Interspersed throughout the living trees were dozens of stumps. He jumped off Titus and checked a three acre area. Slash was everywhere. Trees cut were ponderosa pine and sugar pine. He knew there was no approved timber sale on his district.

He could see where the logs had been pulled through the forest. It dawned on him what the noise was—a steam donkey used for hauling logs from the woods to a landing.

He was just about to mount Titus and ride north, when a horse and rider approached. Buck yelled and the rider seemed startled.

"Who are you?" he demanded.

"I'm Buck Stonewall, the district ranger for District 3 on the Maahcooatche National Forest. Who are you?"

"I'm Miles Rust. What do you want?" Miles was rather short with a beard and balding head. His hands were large and his arms were muscular. He talked fast, and from the aroma, Buck figured he hadn't taken a bath in several weeks.

"Well, Mr. Rust, do you know anything about the trees being cut down?"

"Yeh! I have a sawmill up there," pointing north, "and I needed some logs to operate."

"Are you aware you're cutting on Forest Reserve land and you need to have a valid timber sale contract to operate?" Buck responded.

"That's what you say. These are public lands and I have a right to use the timber like everyone else."

By this time Miles had dismounted and was talking right into Buck's face.

Buck answered, "Yes, these are public lands but you don't have the right to come in here and remove trees at your own whim. You are in trespass and operating against the law. You are ordered immediately to stop any further activity until the government determines your liability. Understood?"

Buck was calm but Miles was getting madder by the second.

"You know where you can go, Mr. Ranger. There's no way you can stop me by yourself. Go ahead if you want to try."

While in Harmony, Buck had stopped by the sheriff's office and obtained a pair of handcuffs. Kent had urged him to do so, since one never knew what kind of low life might be encountered in the forest. He was glad now that he had paid attention.

He walked toward Titus to retrieve the cuffs in his pack and turned away from Miles for a moment. As he opened his pack he felt a pain at the back of his head and knew he was losing consciousness as he fell to the ground on his back. Buck was out cold. He became aware of something nudging his head. Opening his eyes he saw Titus about a foot away. The horse whinnied. Still woozy from the blow to his head he sat for awhile and looked about. The scene made no sense. There was Titus, Stub and Miles' horse. His attacker was nowhere to be seen.

With his horse and mule lined out Buck held on to the reins of the other horse and rode north. About 500 feet later he saw a dozen men gathered around a motionless body. On closer look he could see it was Miles Rust. Buck identified himself and asked what had happened. The men answered they didn't know. They had heard Miles yell and when he didn't arrive they went looking. They said they were employees of the sawmill. Buck could see a portion of it through the trees. He examined the body and saw that the neck was broken and the forehead was smashed. A large, bloody tree branch was lying next to the corpse.

Buck thought he knew what had happened but he remained silent. He remembered that Rust had looked admirably at the big bay. He thought of the first time he mounted Titus, after being warned by the owner who said the horse would not let anyone on his back. He tried it anyway. Titus had taken off like a bullet and headed for some trees and low lying branches. If he couldn't shake the rider off, he was going to knock the rider off. Buck bailed off and Titus immediately stopped. Through kindness and patience, Buck was finally able to ride the horse. The owner sold Titus at a very reasonable price. After almost four years, the two had bonded as much as any man and animal could.

Buck surmised that after Miles mounted Titus, the horse had gal-

Chapter Two - 1906

loped toward the trees with Rust holding on for dear life. He had been afraid to get off and when the bay aimed for a tree sprouting a large low branch there was only one possible result. The rider had yelled just before his forehead collided with the branch. Rust had either broken his neck from the impact or when he hit the ground. Immediately after losing his load, Titus had returned to his owner and nudged his face. Buck wasn't going to tell anyone about his theory. After all, it was only a theory and he was not a professional investigator.

The men figured that Mr. Rust had walked or ridden by the tree at the exact moment the branch had broken loose and killed him. One of them came forward and introduced himself to Buck as Tim Westgate. He was of average height with dark brown hair and a handsome face. In his early twenties, Tim was a happy person who was worried about other people and their feelings. He was an employee but was concerned about the body and Mrs. Rust.

At Buck's request two of the men lifted Miles' body over the saddle on his horse and tied him down. Tim and Buck agreed they would take the body and break the news to Mrs. Rust.

Before leaving, Buck told all the men that the sawmill was shut down and no more timber was to be felled or logs sawed. They were to leave everything as it was. No one made any protest when it was pointed out that their paychecks weren't going to be issued anymore, anyway.

Arriving at the road, Tim turned east and Buck followed with Stub and the "hearse" horse. The ranger asked if there was a saloon in the area. Tim pointed and said it was about 1,000 feet to the south. A few miles farther and Tim turned left on to a weed covered road. After 500 feet they arrived at a farmhouse and dismounted. Tim knocked and waited. The men introduced themselves. They said they had some bad news and asked if they could come in and talk. It took about 10 minutes to tell Mrs. Rust what had happened. She took the news stoically but broke down when shown her husband's body across the middle of his horse. Buck and Tim offered to dig a grave and construct a wooden coffin. They completed the job in about two hours. The hole was covered after Buck said some comforting words to Mrs. Rust as a eulogy.

Before leaving, Buck explained that the sawmill was on government land and the timber was government timber. There would be an investigation and any final liability would probably be covered all or partly from sale of the mill equipment. He couldn't make that determination. They both said again how sorry they were. Buck didn't say anything about his swollen head or his theory. They left Mrs. Rust with tears streaming down her face.

On the return trip the men rode side by side. Tim asked Buck about working for the government. He was now unemployed and had always intended to check out the job of a ranger. Buck explained the exam that had to be passed, the salary, horse, saddle, feed and several other items that should be considered. He gave the young man the names of Kent and Garth. From the brief time they were acquainted, Buck took a liking to Tim and said he would put in a good word to his bosses. Tim had said that every employee had believed they were working on a legitimate timber sale. Miles had told them so and they accepted it.

Buck asked about the saloon. It had been built in a hurry mainly to quench the thirst of the sawmill workers. Tiny Williams was the proprietor and bartender. The stock was limited to beer, whiskey, gin, rye and rum. It was late in the day when the two men stopped at Tiny's Saloon. No one else was there. Buck figured it was probably due to unemployment which translated to no customers. Tiny's horse was tied up out back and the ranger asked Tim to stay out of sight. The place was like a tumbled down shack. No tables or chairs—just a bar. A large man was waiting as Buck walked in unarmed.

"Howdy stranger, what can I do for you?" Tiny asked.

"Well, first off you can tell me if you own this building and sell liquor in it."

"Of course I do on both counts. You are in Tiny's Saloon."

At that Buck identified himself and stated his purpose. He said the saloon was on Forest Reserve land and would have to be removed pronto. He would give Tiny the benefit of the doubt if he cooperated and tore down the building starting tomorrow morning and remove the liquor. As of then the saloon was closed to business.

As Buck talked, the countenance on Tiny's face changed from one of friendship to one of enmity. After the ranger asked if there were any questions Tiny started to yell.

"You can't come in here and tell me I have to close and tear down this building. You have no authority over me or the land I'm on. So get lost before I really get mad."

"So, you aren't going to cooperate or do what I ask?"

"You're damn right I'm not going to cooperate or do what you ask."

"Well then, since it's late in the day I'll be forced to place you under arrest and take you to jail in Harmony."

"Yeh! You and who else," hollered the big man.

Chapter Two - 1906

"See you tomorrow, Mr. Williams."

When they were out of sight Tim turned to Buck and said, "My gosh, what'll you do?"

"I'll do what I said I would," responded Buck. "By the way Tim, I don't expect him to cooperate, so would you be able to gather some of your coworkers and meet me here at 6 a.m.?"

"Sure thing, Mr. Stonewall. I'll have several of them here."

"One more thing, if you have a phone I need to call the sheriff in Harmony and explain the mill incident and to be ready for a new visitor by tomorrow evening."

Tim lived alone nearby and Buck made the call. The deputy asked if any help was needed. Buck said no and as he turned to go, Tim asked him if he would like to stay over for supper and stay the night. Buck accepted the meal invitation but said he would sleep outside. They talked more about the Forest Service that evening. Before lights out, Tim had phoned three of his coworkers and asked them to be at Tiny's by 6 a.m.

Promptly at 6 a.m. there were six men at Tiny's—Buck, Tim and three of his friends. "What's your decision, Tiny?" asked Buck.

"Nothing's changed from yesterday ranger, so get going."

"Well, if that's what you want, then these four men are here to tear your building down."

With that Tiny lunged at Buck. Again it was the kick in the crotch and smashed neck that brought the dispute to a quick finish. It took all four of the sawmill workers to lift Tiny on his horse after Buck had slapped on the cuffs behind his back. When Tiny came to he was in the saddle. He cussed and yelled but it didn't do any good. The men had already started to dismantle the bar. They asked the ranger about the liquor. His reply was that the government couldn't pay them for their work but the government also wasn't going to take an inventory of personal property. As far as Buck was concerned, the liquor didn't exist, so he couldn't tell them what to do with it. They were to pile the lumber in a neat stack. It was lumber purchased from the mill, so it too would be part of the investigation.

When Tiny understood he would be riding all the way to Harmony with his hands cuffed behind his back, he went wild. Buck said he was given a decision to make and he had made the wrong one. They rode south on the main road to Shadowcreek, mile after mile and hour after hour. Tiny refused all offers of water or food. He was still in a foul mood. Buck encountered several ranchers he had visited the year before. He waved and said he would be seeing them soon. The three animal, two man parade made quite a sight.

It was the middle of the afternoon when Shadowcreek came into view. They didn't stop but several people who knew Buck watched in silence. Faye Wadsworth, a good friend of Edna's, was a witness. She immediately headed for the post office.

After turning east from the village of Fish Cut, Tiny finally asked in a nice way if Buck would please change the handcuffs from his back to his front. He explained the pain was killing him. Buck complied and with a couple of miles to go, Tiny kicked his horse and took off. Buck immediately left Stub on the road and spoke to Titus. Within seconds they had caught up with the other horse. Buck took his rope, twirled it at Tiny, lassoed him from behind and pulled him off his horse. The victim was now in terrible shape. He was unable to mount his horse and he could barely walk. Buck rounded up Stub and rode behind a slow moving former saloon owner. Within an hour they entered town and the jail. The sheriff was waiting. Someone had called him from Shadowcreek.

After some paperwork, Buck took all three animals to a livery stable. It was almost dark so he headed for the boarding house and rented a room. After a healthy meal he returned to his room and a long overdue deep sleep.

Rumors flew the next morning in both towns. Somehow, folks had heard about the timber trespass and Miles' death, as well as the removal of the saloon and Tiny's trip. Some thought Buck had killed Miles in a fair fight. Some couldn't believe that one ranger could take care of both problems without help from the sheriff or other law enforcement folks. When Buck was asked about it all from the local newspaper, he shrugged it off as being part of his job. No, he wasn't worried nor did he think he was ever in real danger. He glossed over the details. It all sounded so easy and simple.

When he reported to the supervisor's office, Kent was not in, so Buck gave Garth a blow by blow account of the series of events, except for his theory about Titus. The deputy supervisor listened intently and when Buck was through he let out a big whistle. The ranger said they could expect a visit from Tim Westgate to see if he could be added to an upcoming roster to take the exam. Buck thought the first thing he should do was go to Lakefield Station. Then he would visit the ranchers, check for a phone route and talk to a few more of the folks who lived in the forest. Garth agreed and Buck left. In the meantime Garth would talk to

Chapter Two - 1906

Washington and ask them to send an investigation team for the timber trespass and sawmill situation.

Buck needed to replenish some food and other supplies for Lakefield, so he loaded up at the Shadowcreek mercantile store and headed west. Before leaving town he walked into the post office. Edna was there but so were several customers. She smiled but went about her business.

In 15 minutes it was noon, so the post office closed for an hour. Edna would fix them both lunch at her house while he would tell her the account of his past few days. She had heard rumors but didn't pay attention to such things. Again Buck told her everything that he told Garth. As he spoke, she looked at him with loving admiration. They hugged goodbye and Buck left town for his home in the woods, whistling and singing as he rode.

Lakefield Ranger Station loomed into view. He saw where a couple of the corral fences needed mending from the snow load. The barn was in good shape and the feed was still dry. He unloaded and took care of Titus and Stub before checking on the cabin.

When he did check, the only problem was mice had left tell-tale signs. He would dispatch those little critters in no time with his tried and proven method. There was plenty of dry wood inside for a warm fire.

* * *

Regulations and instructions issued by the Department in a publication dated July 1, 1905, notified livestock owners that a fee would be imposed on and after January 1, 1906, for all classes of livestock. Mr. Pinchot wanted to make livestock folks using the reserves pay a nominal fee for the benefits they were receiving. This caused an angry reaction from stockmen and Western congressmen.

The hostility of the local population for the need to obtain a permit for activities they had been doing for several generations without hindrance was significant. The regulations had been imposed with practically no notice.

Buck had his work cut out for him, explaining the fee and permit system for grazing private livestock on government land. For the next month and a half he rode all over his district. The ranchers welcomed him into their homes and always asked questions about the infamous sawmill and saloon on government land. He became a little tired of answering the same questions but figured it was part of his duties to be a public servant as he continued to be pleasant with his answers. He had no major trouble with the cattle ranchers on his district. He knew their language and their problems. He had worked and lived on several large spreads throughout the West. When he explained the reasons for limiting animals in certain areas and why there was a charge, when for many years there hadn't been any, he heard moans and groans but they were always mingled with the comments that they knew it wasn't Buck's fault. He was just the messenger. As yet there were no sheep ranchers on his district. Buck had heard tales of sheep and cattle wars and he considered himself lucky.

Whenever Buck stopped at a ranch house near noon or supper, he would be invited to stay for a meal. Buck was getting a fond reputation as a story teller. Invariably when he visited a ranch with children near suppertime, there would be pleading and begging to hear some of his adventure stories. Buck always complied, except he had to warn the little ones when his last story for the evening was to begin. His tales were basically true: about rodeos, ranches, gold, travel, wild and domestic animals, hunting and fishing. He would gesture with his body, sometimes with great animation. Alaskan tales were the favorite. He always tried to get some history into the narrative.

* * *

It was April and examination time was near. Buck wasn't worried. He was philosophical. If he passed—fine; if not, he would move on to something else. That was how he had lived but this time there was something different—Edna. He didn't want to leave the area, no matter what. Owners of three of the largest ranches had offered him a ranch manager job if he ever left the Forest Service. He was honored but told them he was happy with his present situation. Life was good and he was content.

Kent had asked Buck to come to his office the day before the examination began. Both of them, along with Nate Bennett arrived at 8 a.m. The supervisor wanted to talk with Nate first so he handed Buck the sawmill trespass report. He immediately turned to the last page for the decision. There it was: The timber cut stumpage value was several thousand dollars. All the equipment at the mill was confiscated by the government and was to be sold at auction with the proceeds to cover the value. If that wasn't enough, then the Rust's other properties were to be seized. Buck read the whole report. It was cold and impersonal. He wondered about what would happen to Mrs. Rust and planned to visit her sometime.

Kent and Nate were still behind closed doors. Lucy had just finished typing a half dozen letters and motioned for Buck to sit down. She asked

Chapter Two - 1906

what he had been doing and how he enjoyed the ranger job. He answered and then asked her how difficult it was to learn to type. He had always wanted to try. She responded that she could teach him to place his hands in the proper position and then it was up to him to practice. It took time and patience and maybe someday he would be detailed to the office long enough to at least start. The door to the supervisor's office opened and Buck was asked to enter.

Kent spoke. "Well men, tomorrow and the next day you will be two of 15 people taking the ranger examination. Two of those on the original list backed out. For your information Buck, Tim Westgate did come to see me and I was duly impressed. I've received permission to add him to the list. First you will take a written test and then a field test. The examiners will explain everything in detail to the group. I believe both of you can pass. Any questions?"

"OK, then for the rest of the day you may read government regulations, policies and statutes plus look at any atlas or record. Lucy will help you find what you want. Good luck, men."

Nate read until noon and didn't come back. Buck read all day and experienced a good night's rest.

The first part of the no-nonsense examination was a written test held in the Harmony Lodge. All 15 applicants showed. Buck and Tim greeted each other. There were two inspectors and one forest supervisor. Men appearing for the examination came from all walks of life.

A statement issued by the Washington, DC office of the Forest Service was read, "Rangers are appointed only after civil service examinations. They must be residents of the state or territory in which the forest is situated and between the ages of 21 and 40. The examinations are usually held once a year. They are very practical examinations. The life a man has led, what is his actual training and experience in rough, outdoor work in the West, counts far more than anything else. Lumbermen, stockmen, cowboys, miners and the like are the kind wanted. It is the hardest kind of physical work from beginning to end. It is not a job for those seeking wealth or light outdoor work."

After formal introductions the test was distributed from sealed envelopes. It lasted all day and its purpose was to eliminate the illiterate. A few of the men's minds went blank. When time was called, one of them handed in his paper with nothing on it. Sample questions went something like:

How many links in a chain?
What is the length of a chain in feet?

How many chains in a mile?
How many feet in a mile?
What is a cord of wood?
What defects are considered in scaling and how would you make allowance for such defects?
If you were allotting a range in a mountainous country, what kind of a range would you allot to cattle and what kind for sheep? Give your reasons.
If you were lost in a rough mountainous country, with which you are not familiar and had no compass, what steps would you take to find your way out?
If you received an appointment as forest ranger in a district where the sentiment was antagonistic to the Forest Reserve, how would you proceed?

The next day were the field tests. Candidates were to locate and describe all four corners of a designated 20 acres, plus make a map showing the topography and approximate amount in board feet of each specie of timber in the area. Next they were required to saddle and ride a horse through several gates and to demonstrate putting on hobbles and picketing with rope. They were rated on time and efficiency.

After lunch the candidates were required to pack a camp outfit of blankets, cooking utensils, tools and barley and demonstrate the ability to load it on a mule. Strewn around on the ground were the pack items. They could use any of four hitches: "diamond," "square," "double diamond," or "squaw." After a trot down the trail and back, the examiner checked the pack to see if it was properly balanced, had not loosened and whether the contents were arranged so as to not damage the animal or the rest of the pack. Those candidates who had never loaded a horse or mule were out of their element. They didn't rate very high with the judges. This part of the test was a real circus.

They were required to set up a compass and run a line on a prescribed course. They had to ride a horse into the woods until they found a blazed line, then follow the line until they reached a stake with special markings.

Next came the use of an ax. Each was assigned a tree to chop down, limb and pile the slash where it could be burned without damage to the remaining stand. They were to chop notches in the trunk to indicate proper log lengths and give an estimate of the number of board feet it contained.

Chapter Two - 1906

The last test of the day was shooting at targets with a rifle at 100 yards and pistol at 50 yards.

Before the group disbanded, one of the instructors gave a talk on forestry practices, forest fire prevention and protection, watershed protection and public relations. It was explained that 70 was a passing score but no one should feel badly if they didn't make it. Each one would receive his score soon.

It had been a long day for everyone. Some who felt they had failed, got drunk at the nearest bar. Others left town. Buck, Nate and Tim ate supper together and talked about the examination. They were glad it was over. Nate told the other two that no matter if he passed, he was seriously thinking of leaving the Forest Service for another job. His wife wasn't happy with his long hours and nights away from home. He liked the work all right but it was hard on his family. He was looking for a more "normal" occupation.

The next morning Buck checked in with the supervisor for any last minute directions. When Kent asked him how he had done on the tests, he replied he probably did poorly on the timber volumes since he had no experience in that area. The supervisor told Buck that Luke would be back on the job May 1st. Buck said he would be at Crescent Ranch on that date to line things out for the two of them during fire season.

Between the exam and May 1st, Buck didn't run into any major problems. It was time to spruce up the cabin at Lakefield. He packed in a nice table and four chairs. He built some shelves for the kitchen utensils, plates, eating ware and food storage. A folding cot for downstairs with extra blankets was added. However, he continued to sleep on the floor of the loft. A flag for the flagpole was requisitioned and purchased.

After taking careful measurements, Buck decided to splurge by bringing running water into the building. The closest water was a swift, running year around stream named Finny Creek. His plan was to erect a small water tank near the creek with water entering at the top and emptying at the bottom. The pipe would be metallic and the end of the pipe would be facing downstream. The water pressure would be by gravity feed from the tank to the cabin. The single faucet would be mounted above a basin and used water would be diverted into the ground outside the building.

He had another idea. Why not add an offshoot pipe attached to the rear wall of the cabin and place a shower head at the end. To date he had used the creek or lake to bathe but the use of soap had bothered him.

This way he could take a long, convenient sudsy shower. After all, the temperature of the water was the same. It took four days of work counting the packing in. Now he could shave, bathe, cook and drink all the water he wanted. Pipe was extended to the corral area for the animal's water trough.

Buck also constructed an ice house in seven days from additional and leftover logs. He planned to haul in some sawdust and during the winter, saw ice blocks from Green Lake. His "freezer" would keep meat and other perishables.

The morning of April 30th, Buck, Stub and Titus left for Crescent Ranch. With an early start he wanted to scout a possible phone line route. Could he go in a straight line instead of winding trail; could the line be strung entirely in trees or were there too many bare spots where poles had to be erected; how difficult would it be to maintain in winter? These were just a few of the questions Buck had to answer in his analysis. He had made the same study earlier between Lakefield and Shadowcreek. He drew routes and distances on sheets of paper as he rode. It was slow going and the sun was setting as he viewed Crescent River and ranch.

With the dogs barking there was no way to arrive unannounced. Four people and two dogs greeted Buck as he rode to the front porch. After 15 minutes with everyone talking, Luke said Buck should have a chance to take care of Titus and the mule.

During dinner Buck talked about his adventures since that fateful day in December. In turn, he wanted to make sure Luke had enough tools and equipment for the coming fire season and he wanted them both to leave for Lakefield the next day. He had something special to show the fire guard. More stories were told that evening. Buck slept near Titus and the next day the two men headed back to the station.

Luke had more experience than Buck in stringing telephone lines, so he was shown the maps drawn the previous day and asked what he thought.

Luke was honestly blunt. "I don't think it is practical at this time, Buck. There must be another route. The distance is long and with sharp elevation drops, it would be difficult to maintain in winter. How does it compare with the Shadowcreek to Lakefield line?"

"It's about the same," replied Buck. "Let's think about it."

The two men rode on in silence. They arrived at the station in the afternoon and tended to the animals. When Luke entered the cabin he

Chapter Two - 1906

couldn't believe his eyes. There was furniture, shelving, a cot—and Buck had the faucet open fully so water was cascading into the sink.

"This is great," exclaimed Luke. "Next thing you'll have a generator."

"Don't think it hasn't crossed my mind," replied Buck.

That evening it turned cold, so with a roaring fire the two men mapped out their district fire patrol plan for the coming months—subject to change at anytime.

May and the first half of June was comparatively routine on the Lakefield District. The usual confrontations with bears, snakes and other wild things occurred. Fire inspections on private land near the forest boundaries went smoothly.

They both happened to be working around the station one afternoon when the sound of horses' hoofs caught their ears. Buck instinctively moved toward his rifle leaning against the wall on the front porch of the cabin. So far, no unsavory people had visited Lakefield, but one never knew. It was wise to be alert.

"Hello Buck! Are you ready for some visitors?" the voice yelled.

Buck relaxed. It was Kent. The other man was Garth. Neither of them had seen the completed Lakefield Station.

The ranger whispered to Luke. "Looks like bosses day at the old corral. Wonder if they're here to make an inspection."

"Glad you guys are here. We just took a chance since the day is gorgeous and we needed to get away from the office."

"It's good to see you two. Been awhile. I'll take your horses to the corral for feed and water. Luke, why don't you get them a drink. If they're off the payroll, then you can offer them some hard stuff." They all laughed but the only request was for water.

After a thorough tour of the compound the supervisor looked at both men.

"You have done a beautiful job. I don't know of anyone else who could have accomplished what you have in such a time frame," said Kent.

"There must be something you wonder about," replied Buck. "Any questions?"

"Well, I am thinking you didn't have funds for your water system. So how did you manage?"

"I thought you might ask about that, Kent. Don't worry, it all came from yours truly. I knew there was no money so I didn't ask. It's for personal convenience, anyway." Buck turned on his jokester look and went on. "That means anybody else using the system must place a dime in the container."

Kent and Garth looked at each other in disbelief. The ranger started to laugh.

"Darn you Buck, you're always doing that and we're always falling for it," exclaimed Garth.

"Of course if you really want to put a dime in the dish it's OK with me," countered Buck. "But please don't put that in your report, if there is a report."

Kent said, "Well, there are really three reasons we're here today. The first was to see the station and the second is to tell you about the recently passed legislation named the Forest Homestead Act of June 11, 1906. All the rangers have to move on this and we thought you might be out of communication for awhile."

Kent went on to explain the details of why, what, who, when and where of the Homestead Act. Forest land users had worked to establish a management policy for the forest reserves from early in its existence. Opponents had questioned the potential lost use of the government grazing and farming lands. These complaints resulted in the 1906 law. It allowed individuals who were occupying the land to claim up to 160 acres of forest land for agricultural use. When the forest reserves were proclaimed, all further homesteading was barred except for mineral purposes. This meant that those living on the land could remain as squatters on their claims and eventually obtain patent and ownership, provided it could be determined the land was fit for agricultural use.

Kent further explained that land use occupancies that required special privileges had to be located and placed under permit. He said that saloons were forbidden and issuance of special privileges for businesses, other than mining activities conducted on mining claims, were necessary. Each ranger was to locate and withdraw administrative sites.

"This is where you two come in," said Kent looking directly at Buck and Luke. "All claims have to be confirmed. You will have to make surveys and recommend if the land is acceptable for agricultural use. You will also have to do an extensive land classification over your entire district. You won't have to actually do the surveying on these potential farm units but you'll have to sketch all areas on a map."

"What about the current workload?" asked Buck.

"Yes, it will make a huge impact on your work. It will take some years to complete the job and you will run into problems with those whose claims are denied," responded Kent.

Chapter Two - 1906

"What does the forester think about all this?"

"I can tell you that Mr. Pinchot was for it. He believes that to farm land better suited for agriculture is to put the lands in the national reserves to their highest economic use. He said the Forest Service would welcome permanent settlers but would not tolerate land speculators. We can expect a flood of claims within the next few months. On our forest we'll have to push the survey work ahead of timber and grazing administration. We're going to try to hire more people but don't count on it." After Kent finished there was a long pause.

Buck spoke. "I'd better get the telephone line survey done right away. We definitely need better communication with the outside world."

"Sounds like an excellent idea," said Kent. "There's a good chance we'll get money for it after July 1st."

The two supervisors indicated it was time to go.

"Wait a second," Buck exclaimed, "I thought you said there were three reasons you came here."

A sly smile crossed Kent's face. "What do you think Garth, shall we tell him?"

"Why not give it to him halfway," said Garth. "Tell him we received the results of the ranger exam."

"Yeh! That should work. We received the results of the exam Buck. That's our third reason for being here. Take care of yourself. We'll be seeing you."

"What!" yelled Buck. "Just a darn minute. Boss or no boss, you can't just come up here and tell me that without the rest of it. What happened?"

Kent and Garth were mounted and waved to Buck and Luke as they started to return home. Buck ran around to the front of the animals and grabbed both halters. By this time the riders were laughing.

"OK, you guys got me this time," said Buck. "Guess I deserved it. Now please don't hold me in suspense any longer. Did I pass or not?"

"Well," replied Kent, "I hate to be bearer of bad tidings but out of 15 candidates only four passed. You'll be glad to hear that Nate Bennett and Tim Westgate were two of those. I don't remember the name of a third one. As to the fourth . . . congratulations, Buck. You made the highest score. It was in the 90s but I don't remember the exact figure."

"I can't believe it," cried Buck. "I had problems with the timber and volume part and my pistol marksmanship wasn't very good. Thanks, you two, for coming up and letting me know. Guess you won't be firing me after all."

"Congratulations, Buck" Garth spoke as he held out his hand. "We were very relieved with the result. No, you can't get out of the job anymore. You're stuck with us—and we're stuck with you."

Buck let go and both horses trotted off. It would be dark by the time they arrived in Harmony.

Luke also congratulated Buck and said how much he enjoyed working with him during the short time they had known each other. "Now what are you going to do with that telephone survey?" he asked.

"I've been thinking about it from every angle," replied Buck. "What if we went at it indirectly instead of only two choices from here to your ranch or to Shadowcreek?"

"What do you mean?"

"Well, the closest line by a direct route is from here directly south to a point on your line to town. It would tie in about halfway between Crescent Ranch and Shadowcreek, I believe. I haven't ridden the course all the way but I think it would be almost 100% tree line. The distance is shorter than the other two, the elevation changes are gradual and it would be cheaper and faster to install with no poles to erect. What do you think?"

Luke thought for a moment. "I think you've hit on the best idea so far. The other options just didn't seem practical. Remember during the winter, there will be times when the line breaks due to trees falling, snow or ice, and if you want to keep in contact you'll be stuck with the repair and maintenance."

"Yeh! I know all the landowners along the line and I don't think we should have any trouble when they realize it's to their advantage to be able to contact the ranger adjacent to their property for fires, rescue operations and routine questions such as how are the fish biting. Let's figure on getting a good start in the morning and finishing the project in three days."

They worked around the compound the rest of the day. Next morning they got all the tools and equipment needed, including paper and pencils. It was a fairly direct route but they did have to adjust for the lack of trees in some places and outcroppings of rocks here and there. They crossed several streams. Flagging was used to mark the route.

The private land they entered was owned by Peter Blodgett, a widower in his mid 70s. Luke had known him for years. He was a good man and several times had mentioned to Luke he was getting too old to do the work and keep the place presentable. They estimated that about six poles

Chapter Two - 1906

would have to be installed between the last tree and the splicing into the main line along the Crescent River. One of the streams crossed by the line was Music Creek. It flowed through Blodgett's small ranch on the way to the river.

The two men knocked on Mr. Blodgett's door. Luke introduced Buck and Peter invited them in. The old gentleman was gray and gaunt. His mind was sharp and his eyes sparkled as he talked. The years had taken a toll on his wrinkled skin and stooped back.

They talked for some time but finally Buck got around to the telephone line project. After explaining the project and their desire to obtain permission and an easement for a line through his property plus installing the necessary poles, Peter thought for some time and then answered.

"You know, I've been thinking about selling this place within the next four or five years so I personally don't care about the lines and poles in the long run. But they do need to be placed along the property line so it won't interfere with whatever the new owner wants to do with the land. I also agree it would benefit the locals along the river to be able to contact the ranger in charge. So with that one limitation I'd be willing to allow the phone and sign the easement."

The two men thanked their host. Nothing was for sure until they received permission and the funds. If those things went well, then Mr. Blodgett could expect a line to be installed sometime before winter. They would let him know either way.

Buck took more measurements. It was time to put it all together and present the plan to Kent. The ranger wanted his analysis and presentation completed as perfectly as he knew how. His plan was to go to the supervisor's office and spend a day or so making maps, measurements, time schedules and cost estimates for materials and labor on all three routes. Luke headed west for home and Buck west toward town. They agreed to meet at Lakefield station in four days.

The post office was closed by the time Buck rode into Shadowcreek. He didn't want to see Edna until after taking a bath and eating. Faye Wadsworth at the Pilot Hotel was glad to see him. She was a kidder and could make personal comments to her guests without fear of an unwanted response.

"When are you and Edna going to get hitched?" she blurted.

Buck was completely taken aback with this question.

"Don't be silly, Faye. What a ridiculous question. I haven't thought about it and Edna has never come close to suggesting it. You shouldn't kid about something like that."

"Now don't deny you've never thought about it. I've seen you two together. The world could be coming to an end and neither of you would budge when you talk and look at each other the way you do. I've been around Buck and I know real love when I see it."

Buck didn't know what or how to answer. Finally he said, "I'll admit over the years I've enjoyed the friendship of several women and I've always left town. I'll also admit that since I've known Edna, I've never thought about leaving. But I would never ask her or anyone else to be my partner for life and live in isolated wild and primitive conditions like Lakefield Station. So you see there is no need to even think about it. Let's drop the subject. Anyway, I need a room."

Faye was smart enough not to push it and she surmised Buck wouldn't say more about the subject.

After a bath, a shave, and a quick bite of food, Buck walked to Edna's house. He had previously left Titus and Stub with Elroy at the livery stable. Before he even knocked, the barking of her golden retriever Molly was heard. It was late but Edna had told him to come at any hour, so he knocked. Discovering it was Buck she opened the door wearing a bright colored wrapper[6] and invited him in. After talking an hour to catch up, the ranger looked at the clock on the wall and said he had better be going. Neither wanted to part but as levelheaded adults, they knew it must be. Edna asked if Buck would be in town on July 4th to watch the fireworks again. The fourth was on a Monday so she had Saturday afternoon off and all of the next two days. It would be a long weekend. Buck said he hoped to make it and left after a long embrace.

As always, Lucy arrived at the office promptly at 8 a.m. Buck was waiting and asked if there was a desk he could use to put together his "communication" plan for the Lakefield District. He was engrossed in his work when Kent walked in and saw what he was doing.

"Gee Buck, when you say you'll do something, there's no wasted motion."

"Yeh, boss, I'll be working on this for a day or so. When you receive the report it will cover everything you need to make a decision, including three alternatives with a recommendation for the best one and detailed maps."

"OK, then I'll leave you alone."

For that day and half of the next, Buck kept at it without a break. He asked Lucy to type part of it. By the time he talked to Kent the conclusions were self evident. The only question the supervisor had was could

Chapter Two - 1906

Buck and Luke do all the work alone. Buck said they might need help on the pole installation.

There had been no requests or claims to check any homestead applications so far on Buck's district. The other three districts already had some and the rangers were busy surveying. Kent had one more thing he wanted Buck to listen to before leaving.

"I've just received Nate Bennett's resignation as ranger on District 1. As you know, his district is the smallest on the forest and is the closest to Harmony. It has a permanently built station with separate residences and a bunch of amenities compared to your district. I want to give you a chance to move there as ranger if you want to. Think about it."

Kent went on to explain that the District 1 station had previously been the headquarters of a logging camp. The company had gone out of business at the turn of the century and the General Land Office in the Department of Interior had purchased the improvements at an extremely low price. During the past few years it had become a well developed station which was unusual at that time.

"That's really great to think of me, Kent. Ordinarily I might be interested but I can tell you right now that I'm very happy where I am. Lakefield District is the biggest challenge in my life. Before leaving it, I hope to have roads, phones, residences, campgrounds, bridges and much more built. We need lookouts for fire and more guard stations, as well as lining out timber sales and analyzing grazing districts.[7] Feel free to ask someone else. Thanks again for offering it. I'm flattered."

"I thought you'd say something like that, Buck. I really do want you at Lakefield with all the work the district needs. Since Tim Westgate passed the exam, what do you think about him as ranger on District 1?"

"An excellent choice. I feel you could not do any better. It would be great having him for a neighbor."

"I'll let you know if we get the money for your phone line, Buck. You probably want to get back so that's it for now."

They shook hands and Buck headed for Shadowcreek. He had a proposal for Edna. One that he had thought about since her long weekend statement. The post office was open. He was glad since Edna had mentioned there was a letter from Nebraska. With one other person inside, they kept their conversation formal and businesslike. Edna handed Buck his letter and he tore it open. Ivy was expecting. Buck yelled and said he was going to be an uncle. Edna and the customer congratulated him.

Buck asked if he could buy Edna's dinner. He had something to ask that was important. He needed her consent on a proposal that was somewhat daring and possibly improper.

Conversing as they ate at Mabel's Café, Buck told Edna what was on his mind. "I got to thinking about the fourth of July weekend. Since you've never been to Lakefield Station and I haven't shown you any of my district, would you be willing to ride up with me to Lakefield on Saturday afternoon? It would be too late to go back the same day so you could stay overnight downstairs on the cot and I would stay either upstairs on the loft or in the barn. Molly could come along and you could bring any firearm you wanted. There, I've been rehearsing this speech for a day and I've said it. If you think the idea is silly or I'm too bold to suggest it, please tell me."

Without moving Edna was looking straight at Buck as he talked. She smiled and let out a little laugh when the speech ended. "You know Buck, I can tell when you're nervous about something. I knew when you came into the post office today that you had something important on your mind. I'm getting to know you and your ways. But to answer your proposal—yes, I'd be happy to see Lakefield with you and stay overnight. I don't expect you to sleep in the barn, however. But I do expect to bring some food and fix our dinner—agreed?"

Buck was ecstatic. They laid plans for what food to buy and the rest of the logistics. They both would live in anticipation of their adventure. It was as if they were teenagers again.

Buck returned to Lakefield and Luke showed on the agreed day. They went patrolling for fires.

July 4th was still a few weeks away. Buck wanted to spend the time improving as many trail miles as possible. The trail system on the district was poor. The tread was overgrown, full of rocks, holes and often nonexistent. Tree branches were too low and the route sometimes needed to be redirected. So with shovels, double bitted ax, saws, brush hook, mattocks, crowbars and other tools and equipment, Buck and Luke spent many days on the betterment of the trail system. The route from Lakefield Station to Shadowcreek was the most traveled so they completed it first. Next came the trail from Lakefield to Crescent Ranch. Finally, trails to the top of Tip Top Mountain and Florin Peak were altered. Still there were many miles to improve. A trail from Lakefield to Barnesville Ranger Station on District 4 was in poor shape. The trail that Buck had taken earlier from Shadowcreek north on the west side of Shadow Creek needed attention, as did one along the west side of District 3 from Crescent

Chapter Two - 1906

River to the Hondo Indian Reservation. These would have to wait. It was Saturday July 2nd and Buck wanted to get a pre-dawn start to town. He bid goodbye to Luke for a few days as the guard headed for his ranch and fire station.

Buck's spirits were high and he made good time on the "new" trail. It was Saturday so he didn't call the supervisor's office. Edna worked until noon and was ready to go with the food and other supplies. All she had to do was pick up her horse from Elroy. After a quick lunch they were on their way to Lakefield by 1:30 p.m. As they traveled, Buck pointed out various land features. They crossed several streams and saw many critters and birds. Molly was having a great time. It was a glorious ride. Approaching Lakefield Station Buck pointed out the location where the yellow jackets had attacked the pack horses. He showed Edna where they had stacked the logs for the cabin and barn and the encounter with the bear.

After unpacking and placing all three animals in the corral, the guest of honor was given a detailed account of how the cabin was built. Buck showed everything—the water system, shower, privy, corral, barn and the inside of the cabin with all the furniture, kitchen area, loft and shelves of food, posters and official paperwork, plus all the fire tools, hand tools, piles of wood and feed and tack for the animals.

Edna marveled at it all. It was time for supper. She made an excellent beef stew. The bread had been baked earlier but was fresh and delicious. The leftovers for Molly were sparse.

It was too warm to have a fire so they brought out two chairs and sat on the porch. Since neither of them used tobacco, there were no interruptions to their watching, listening, smelling and contemplating the evening sounds and odors of the forest. A soft rustle of branches and hardwood leaves testified to the slight breeze. It was a delightful time.

Edna broke the silence by asking Buck why he didn't have a dog. She said he was about the only outdoorsman she had ever known that didn't. His response was that he had dogs as he was growing up in Nebraska and he liked them. After leaving home, he claimed, a dog became an extra problem as he traveled and he didn't want to get attached to one.

Edna accepted the explanation and then asked if he would like to have one now. Buck thought he might be ready but it would have to be a puppy so he could train it as it grew. The subject was dropped and the wonders of mother nature were discussed as the full moon rose.

Chapter Two - 1906

Suddenly Molly started barking violently. Buck instinctively grabbed his rifle standing by the door. Edna retrieved hers from inside the cabin. Looking in the direction Molly was barking, they saw a mountain lion approach. They were between the lion and the barn where the three animals were kept. It crept slowly toward the two humans. At fifty yards Buck took aim and fired. He missed and let out a rare swear word. At the shot the cougar bounded toward them. Before Buck was ready to fire again the animal was almost on him. A shot rang out and the cat fell dead, shot directly between the eyes. It took a split second for Buck to realize Edna had fired the fatal shot. For some time he praised her marksmanship and courage. She accepted the compliments but didn't mention her past as a crack shot.

The lion was a young male. Buck couldn't understand why he had stalked humans. There was plenty of game in the forest and he had never before seen a sign of cats around Lakefield.

Hesitantly Edna replied that it was probably because she was there. Buck didn't understand her meaning.

She said, "You know Buck. I'm in that situation right now and the scent probably reached the animal."

"Oh, yeh!" Buck replied. "You may be right." He silently admired her for having the gumption to tell him. What an amazing gal, he thought.

Buck buried the cat at a spot outside the compound area because he didn't want the carcass to attract other wild animals. He wasn't interested in the $10 bounty paid on all cougars.

It was getting late but Edna had started to sing "Come Where My Love Lies Dreaming." She had a beautiful voice and Buck didn't want to ruin it with a duet so pulling out his harmonica he accompanied her until the end. They both laughed after Edna said she didn't know Buck played the instrument so well and in return he hadn't realized she sang so beautifully. Another half hour was passed with music filling the nooks and crannies of the nearby woods.

Before Buck climbed the ladder to his loft and Edna prepared for her cot, they both hugged and agreed it had been a magical evening. Molly slept just inside the front door.

The next day was spent at Green Lake. The wild flowers were still in bloom and the bees were active. They had planned for a picnic lunch. A blanket was laid down on a peninsula where the entire lake could be viewed. Squirrels, chipmunks and ants tried their best to be invited guests. After lunch they fished. Edna caught the largest. Buck caught the most. Heading back to the station Buck commented that he wished Edna could stay the night. She replied she had intended to, unless Buck had other plans. They ate a delicious rainbow trout meal.

Monday about 11 a.m. Luke came riding through. He had received a telephone call about a fire on the northeast side of the forest. Buck was ready with Titus and tools in 10 minutes. He made sure Edna would be all right going home on her own. He had seen her use a rifle and wasn't that worried. She bade him farewell and asked if she could borrow Stub for a few days.

The two men took off and about an hour later the postmistress left the station with her horse, Stub and Molly. She would watch the fireworks that evening without Buck.

Titus and Baron took their passengers to the far northeast corner of the original reserve. When they came to Shadow Creek the smoke was about a mile away to the east. Sizing up the situation they both agreed it would take too long to go north then east then south to the fire. The creek was running high. Buck secured his hat and whispered into Titus' ear. The bay leapt forward into the rushing liquid. It was a tense moment. The horse hit with such impact that his body was completely covered with water. He managed to keep his head in the air but there was no bottom. They had entered at a deep hole. About 200 feet downstream they emerged from the river, wet but happy to be alive. Next it was Luke's and Baron's turn. By going to school watching the first two, they moved south to a shallower entrance. Baron jumped in and had a much easier time than Titus. After emptying their boots and checking their tools, the wet little group headed directly for the fire.

It seemed to be about 30 acres. Luckily there was no wind. Both men tied up their horses at a safe distance and immediately started to build a line. It would be a long fight. The heat of the fire helped dry their clothes but it would be a couple of days before the dampness completely disappeared. They worked on containment.

Later several ranchers known to Buck showed up and helped. They were just in time. The two men hadn't slept in three days and nights. By Friday evening the fire was controlled. Luke would stay with two ranchers to watch for hot spots for a couple more days. Buck went back via the main road to Shadowcreek Saturday morning, July 9th. He was a filthy mess and Titus wasn't much better off. He certainly didn't want Edna to see him this way, so he rode west from the outskirts of town directly to Lakefield Station.

As Buck viewed the compound there was something amiss. Stub and

Chapter Two - 1906

a strange horse were in the corral and the cabin door was wide open. He was tempted to draw his rifle but thought better of it.

"Hello, who's there?"

Edna ran out of the doorway and yelled, "Hi Buck, I'm glad you're here. I wanted to wish you a happy 30th birthday."

Buck was speechless. He hadn't even remembered his birthday, let alone the idea that Edna would be there to greet him. "I'm sure glad you're here but I'm such a dirty sight that I won't even come into the cabin until I've taken a shower. Would you please get me some clean clothing? I need everything!"

Edna disappeared and Buck led Titus to the barn and a good wash. The bay went from a starvation diet for five days to all he could eat.

Buck trudged to the shower. He started to peel before Edna brought his clean change. She rounded the corner and apologized, set his clothes down on an outside table and withdrew around the same corner. As the drain water passed by it was black. Buck was moaning with delight even if the water was cold and took some time for the creation of any lather.

It was a comical scene. They were each three feet from the opposite sides of the same corner of the building. Buck was buck naked and Edna was in a new riding outfit. They acted like an old married couple. As soon as the shower stopped, she reached around the corner with her right arm and held out a towel. Buck thanked her and vigorously rubbed himself dry. Five minutes later he was dressed and planned to shave inside. Edna came around the corner and they hugged. She then told Buck to shut his eyes before he went into the cabin. On opening his eyes, there before him was a new cambric[8] tablecloth, new chintz[9] curtains hanging from rods, a pair of great looking candle holders with candles lit next to a pound cake with three candles (one for each decade) and a first aid kit on the kitchen table. She sang a happy birthday song and he stood with his mouth open. Regaining his senses Buck strode over to Edna, held her in his arms and for the first time kissed her tenderly but passionately on the lips.

That evening Edna noticed Buck had several bruises from the fire. She took a bottle of arnica[10] from the kit and placed the tincture on the sore spots.

Edna explained that she hadn't known when Buck would show up, but she was going to stay the night and the next day. At the last moment she would write him a birthday note and leave for town. When she heard him approach, all her presents were in place and she lit the two large candles. The only cake she could bake at home which would hold up for a four and a half hour ride was a pound cake. She had noticed there were no first aid supplies so she made one up and included a little of everything. The tablecloth and curtains were to brighten up the quarters. She hoped he liked the colors. He couldn't stop thanking her enough. They sang and talked that evening on the porch.

Next morning Buck's rifle went through a thorough cleaning. Edna scrubbed his clothes on a washboard and hung them to dry. After lunch they buttoned up the compound and headed back to Shadowcreek. Buck needed to make a report on the fire and find out what had been going on at the office.

Kent had some great news. Buck now had the money to start the telephone line from Lakefield to Crescent River. He had approved the analysis and immediately sent the request to the Washington office. His request had been so thorough and logical and timely that they couldn't refuse. Buck was delighted.

Kent also handed Buck a new *Use Book* dated July 1, 1906. It was the same dimensions as the original one but had increased in size to 208 pages, including the appendix and index. It said all previous regulations in conflict with the current ones were hereby revoked.

Buck knew the Shadowcreek Mercantile didn't have the necessary inventory on hand so he immediately called Calvin Nibbs and ordered a large amount of supplies and equipment. It consisted of miles of No. 12 iron (galvanized wire), hundreds of solid insulators and brackets, climbing hooks and rope, hand crank phone, insulated pair of wires, two lightning protectors, a box (fuse box), six poles and other special tools and equipment. The wire should be in quarter mile rolls. Buck explained it was an urgent order and he would appreciate any special consideration. Calvin said he had some of the items but even with fast delivery it would take at least three weeks. Buck understood and responded by saying he would pay for any added cost to speed the action. Delivery would come by train.

Buck contacted Luke and said they were in the telephone construction business. The guard was to continue patrolling the district but at the end of three weeks he was to contact the supervisor's office. The ranger reserved enough pack horses from Elroy for the project.

Buck worked around the Lakefield compound for the three weeks. He completed the ice house by placing boards over both sides of the logs. This increased the insulation effect. The one door was two and a half feet wide and four and a half feet high. He still needed to store a large quantity of sawdust in the ice house before winter. The building was a cube shape with each of the three inside dimensions at seven feet. He figured to place the ice exactly six feet in all directions which would allow six inches of sawdust between the ice and each of the six sides. He brought the four yards of sawdust needed on a long pack train.

When the three weeks were up, Buck left for town. Calvin said the phone line and equipment had arrived the previous day. The ranger was elated. He called Luke and asked him to meet the next day in Shadowcreek. The guard should stop at Peter Blodgett's place and inform him of the news.

The pack train arrived at Lakefield Station two days later. Every animal was fully loaded. A pair of wires was run from the outside box to the crank phone in the cabin. One of the lightning protectors was installed. The other would be placed where it connected with the commercial line. From the box the single wire No. 12 ground return circuit was connected.

The two men headed directly south blazing a primitive trail as they traveled. Luke had more experience in climbing, so the first few days he used the tree hooks and climbing rope. If branches needed removing they were cut down with a hatchet. The wire was hung on solid insulators spiked to trees. The problem was that every time a tree fell across the line, it broke and frequently tore off the insulator.

The distance between trees averaged 50 feet. At the end of each quarter mile of line a Western Union splice was used.

Eventually Buck climbed and Luke stayed on the ground. Although they had to find a proper tree at regular intervals, the woods were so heavily forested the route was almost directly south. The weather cooperated and as the days progressed they returned to the station each night. At the halfway point, Buck decided they were wasting too much time traveling, so gathering food, tents and supplies, they rode to the three-quarters point and erected a base camp.

Almost two months had passed. They had gone to town for a month's food supply once. They were now within hailing distance of Mr.

Blodgett's ranch. The six poles had been delivered. Peter was glad to see the ranger and sign the easement. Buck showed him exactly where each pole would be placed.

It had been some time since the latest contact with the supervisor's office. Garth told Buck there were two claims made under the Homestead Act that needed immediate attention. He could pick up the necessary equipment such as compass, chain (66 feet), scribe used for marking corners for township and range and Jacob (Jake) staff.[11] Also available at the office were such supplies as pencils and note books. After obtaining the claim locations, Buck and Luke headed for Harmony.

Measuring claims and making determinations was only one part of the Homestead Act. Buck knew they had to inventory the resources of his district under an extensive land classification survey. This meant locating and sketching any area that was large and flat enough that it could possibly be considered a potential farm unit. Most of the land would be along creeks and river bottoms in scattered tracts. It was just another project that needed doing but Buck had made the decision to prioritize trail work, phone line installation and other improvements.

Luke and Buck got right on the homestead claims. Since the land had previously been surveyed by the General Land Office, regular geometrically rigid land descriptions were used. Otherwise, property lines could conform to topographical features described by metes and bounds survey. One claim was in the southeast portion of the district. The other was in the rugged Western area in higher elevations. It was in an area where the soil was poor for growing. That coupled with the difficulty in clearing such a large area, made it unfit for an agriculture classification. The claimant threatened Buck initially but the ranger persisted by talking common sense and pointing out the likely possibility that he could work hard for five years and then have to abandon his "stump ranch" for lack of a crop due to the short growing season and no roads for transportation to market.

All Homestead Act reports required a long list of items to be covered: township, range, section, subdivision, surveyed or not, name of claimant, married, single or divorced, if female and with prefix Mrs. or Miss, elevation, soil, grass or timber, number of acres in each; board feet of timber, climate, rain and snowfall in inches, number of acres under ditch and cultivation, kind of crops raised, duration of growing season, occurrence of frost, distance from market and means of transportation such as wagon or railroad; power site possibilities; if the tract was more valuable for agriculture than timber and some others. The

Chapter Two - 1906

results were mailed from the Crescent Ranch. Approval of any type of homestead claim had to come from higher authority than the ranger.

Buck and Luke returned to work on the phone line. They had lost five days. It took them some time to dig the six holes and to firmly pack the poles into the ground.

A dead end insulator was installed on the telephone company's pole and a jumper wire placed between the Forest Service line and the commercial line to the Shadowcreek exchange. Since the latter line was a ground single wire system, the two lines were compatible. The exchange assigned Lakefield Station one long and three short rings.

Hurrying back to Lakefield the two men cranked the phone instrument and crossed their fingers. They called the exchange and got through. What a relief! The operator was asked to call back to make sure the ringer worked. She did and they picked up after one long and three shorts. There were several other clicks on the line and they knew others were listening. It didn't matter, they now had communication with the outside world. Completion was ahead of the first snowflake.

They both knew that during the winter the line would be broken more than once by falling trees and heavy snows. It meant obtaining some snowshoes for maintenance use. Buck had used shoes before in Alaska but they were new to Luke.

With the good weather holding, Buck and Luke continued improving trail. Completing the trail along Shadow Creek they were halfway on the west side trail before the first snow obliterated the tread. Luke would be off the payroll anytime now. They had been extremely lucky to have had only the one large fire in July. There had been some individual tree strike fires which were controlled within a few days.

Buck contacted the supervisor's office by telephone from Lakefield. Kent said Luke was off for the season starting the next day and Buck was to report as soon as possible to his office. The next morning the two friends shook hands and rode in opposite directions.

The supervisor had been ordered to work in the Washington, DC office as district forester for the winter months. It was something that several supervisors had been detailed to for experience. Garth was to be the acting supervisor and the two of them had decided to have one of the forests' rangers take over Garth's regular duties. Buck was picked. It was to his advantage to gain knowledge of work at the supervisor's level. What did Buck think? He answered that he was flattered and would give it a try but hopefully it would last only for the winter. He had many projects on the district and he wanted to work on them starting in the early spring. Both Kent and Garth were fine with that understanding.

Buck made a long term reservation at the Harmony boarding house. He placed Titus and Stub at the local livery stable and started working in the front with Lucy until Kent moved out. Garth would stay in his own office and Buck would work at the supervisor's desk.

It was October and Buck had something on his mind. What to get Edna for a birthday present? She would be 25 on the 15th of October. He thought about the milliner in town. With so many hats, there was no way he could pick one and make sure Edna liked it. The local dress store also overwhelmed him. What to do? One day walking by the jewelers, a beautiful cream colored cameo with a blue background caught his eye. That was it. He bought it and asked the clerk to wrap it.

When Edna opened the gift she cried with joy, hugged and kissed Buck profusely and told him it was the most wonderful gift he could have given.

Buck worked in the office on atlases, making office repairs and learning to type. This last one was done on his own after hours. By the first of the year he could type about 30 words a minute.

That fall the first Forest Service uniform was made available. It was brown with a green cast. Since there was no requirement to wear it, no one on the Maahcooatche ordered one.

Looking back over the year, Buck was satisfied. It had been fulfilling and a challenge. He looked forward to 1907.

1907

The Ranger's Life

Nights that are spent in the open,
    Under the whispering trees;
Slumber that's sweet and dreamless—
    Lullabys sung by the breeze.
Waked by the first red sunbeam
    Unto no day of strife—
Such is the ranger's life.

Over paths flecked with sunshine,
    Threading the tree-lined ways;
Fording a snow-born streamlet
    There where the big trout plays.
Surprising the elk at the dawning—
    The bear at his clumsy play—
Such is the ranger's day.

Think you the city can call him?
    What charm has the market place?
Why should he turn from the mountains,
    Inviting, from peak to base.
Town's but to dream of at even,
    When camp fire smoke curls high.
So lives the forest ranger
    Under the western sky.

Arthur Chapman

CHAPTER THREE

1907

Two name changes occurred during the year. On March 4th, Forest Reserves were designated National Forests and special privilege permits became special use permits.

Kent came back in April after his Washington, DC detail. He talked to Garth, Lucy and Buck about his positive experience but he was extremely happy to be home.

Garth told Kent the forest received funding for three assistant ranger positions. He wanted the supervisor to make the decision as to where they should be stationed. Buck was asked about using an assistant on the Lakefield District. He thought about it but told his boss to let the other rangers, Tim, Ralph and Hugh have them. If there was any other money he would like to hire another fire guard. Luke's station was in the southwest part of the district and he could use one on the northeast side. Kent agreed and said he would let Buck know about any additional salary money.

Buck made another request. He had been thinking about the feasibility of building locked tool caches, loading them with hand tools such as shovels, axes, mattocks, hazel hoes, brush cutters and picks, and placing them at strategic locations on the district. He pinpointed six spots on the map where he thought they would do the most good. The material for each box and the lock and tools to fill it came to $18.50. He had done his homework. Both Kent and Garth liked the idea. Garth said he had $100 that could be used. This would cover five boxes. Buck said he would cover the extra $11 to make six.

During the winter months, Buck saw Edna on weekends. This meant he usually ate one good home cooked meal each week. It was time to return to Lakefield Station, however. Luke wasn't due to come back for another month. Then the two would build and install the fire cache boxes. They would also do a total analysis of the district's agricultural lands to satisfy the Homestead Act, plus some boundary surveying. The boss had told Buck to offer at least one large timber sale. He was to get help from Hugh Tanner stationed at the Barnesville Ranger Station. The forester wanted better estimates of timber on each forest, so rangers were to make the best estimates possible. Buck thought he could do it all in one summer's work, plus still patrol for fires and trespasses. He would combine the timber and agriculture analysis work as he went. Buck still wanted to try his ice house experiment for Lakefield so he signed out two pairs of snowshoes from the warehouse in back of the office in Harmony and rode for Shadowcreek with Titus and Stub. A toboggan, ice tongs and long crosscut saw were purchased. When Buck arrived at Ike's Livery Stable, Elroy Taylor was busy using a butteris[12] on a horse.

Buck asked if Elroy would knock every other tooth out of the crosscut. Elroy did so and the ranger explained it was now an ice cutter. He had done it successfully in Alaska for cutting ice blocks from lakes. Two pack horses were rented and Buck purchased enough food and supplies to last a month. Edna was working at the post office when he passed by with his four-animal pack train. Stub packed the toboggan, tongs, saw and rope. One horse packed the food and other supplies. The last horse packed the animal feed. Buck couldn't take a chance there was still good feed left at Lakefield.

Edna ran out and waved after Buck whistled. It was late and the ranger knew they wouldn't make the station until dark. The snow was deep in places but the trail was visible most of the time. He was thankful they had improved the trail the previous year. A few times the pack horses stumbled but regained their balance with no loss of cargo. It was slow going.

At long last, ahead in the half moonlight were the outlines of the cabin and barn. The snow had melted enough so the door to the cabin could be opened and a lantern for light was retrieved. The barn was intact and the feed there was still dry. Buck offloaded the three animals and the saddle on Titus. Food and water for all four equines were distributed. The ranger was tired, hungry and cold. Back in the cabin he looked around. Signs of mice were present and some of the wooden chinking boards had been gnawed, probably by porcupines. He had left a good half cord of dry wood in the room. It took a while to get the fire started, but once it had, the place warmed up in a hurry. After placing extra wood on the flames, he lay down in his clothes without taking off his boots and slept soundly.

The next day was clear and cold. Buck tried cranking the telephone. There was no resistance so he knew the line was open with a broken

Chapter Three - 1907

wire somewhere. Grabbing a bite, he headed for the ice house. The sawdust was piled high inside. It took awhile to shovel it outside near the door. He left six inches inside covering the floor.

Saddling Titus, Buck fastened the toboggan with sideboards attached, along with rope, crosscut, ice tongs, hatchet, chisel and hammer. There was almost a foot of snow between the station and Green Lake. Noting the ice was about 12 inches thick where he would cut, Buck made equal blocks of one cubic foot. Using the chisel and hammer or hatchet, he smoothed out the rough edges and bulges on each piece. Twelve blocks were secured to the toboggan and Titus moved downhill. Although there were some difficult maneuvers, they succeeded in repeating the process a couple more times that day.

Buck figured he would need 208 blocks of ice. This allowed a food chamber in the middle of eight cubic feet or eight blocks in size. The first row was placed along the rear of the room at six blocks long and another six blocks high. Before starting the second row, he placed six inches of sawdust between the ice and the back wall. Building each row toward the front, he packed the sawdust in the open spaces. The food freezer was thus surrounded on all sides with double rows of block ice. Boards were placed over the center of the fourth tier as a ceiling for the chamber. As he progressed, Buck poured water over the ice so the blocks would seal together. The access to the food was through eight blocks just inside the door. So they could be easily removed and replaced, he placed straw between the eight to prevent their solidifying together. It took him five days to complete the job. Buck hoped the ice would last at least a year or maybe two.

Before warm weather set in, Buck hauled fresh meat from Shadowcreek to Lakefield. He carefully placed it in the ice house making sure the individual packages were labeled for contents and amount, gleefully visualizing sitting down with Edna and producing two juicy steaks. She would be impressed and he could hardly wait. For the rest of the year there would be meat on the table for all who visited. With such a small compartment, however, it meant replenishing the meat throughout the summer. For a variety there was plenty of fish at Green Lake. Civilization was continuing to encroach on Lakefield Ranger Station.

Buck had originally planned to follow the phone line on snowshoes and repair any breaks. However, the pack horses and Stub were showing effects of the cold, so he left the two pair of snow shoes in the cabin and departed with the whole train the next morning. Halfway to town it started snowing. By the time they got to Ike's, there was a foot on the ground.

Elroy was in the middle of shoeing a horse. He told the ranger he had a half dozen more to go that day and was falling farther behind. Buck said he would help and since Titus was badly in need of some new shoes, he asked if he could do his horse first. Elroy left for some other work and Buck took over the shoeing. He was nearly through by the time the blacksmith returned.

After the last three horses were finished, Elroy asked, "How much do you want for your work, Buck?"

"Why nothing, Elroy. I did it as a friend."

"That's mighty generous of you Buck but I'm not going to charge you for your own horse's shoes."

"You can't survive by doing things like that."

"I'm still surviving Buck and I've got a lot of good friends like you."

Buck laughed and said, "I give up. Thanks very much."

Buck took his leave and headed for the Pilot Hotel. He hurriedly washed up, dressed and strode to Edna's. They hugged and kissed and both talked at the same time. Edna said there was a letter at the post office from Nebraska. It looked like a man's handwriting. Although there was no indication of urgency on the envelope, Buck still worried and talked about it. Finally Edna said, "All right Buck, so you can stop fretting we'll go to the office and open the envelope."

Off they went in the still falling snow. Buck opened the letter. It was from brother Alva. He read it out loud to Edna.

Dear Brother Buck,

    I hope this finds you well. The four of us are fine at the farm. Ivy is a great mother to Charlie who is now two months old. I don't get much sleep so sometimes I stay in the barn. That is not the reason I plan to leave the farm, however. As you know I am now 20. You left home before that age to become independent and see the world. I believe the time has come to do the same. Ivy and Henry tried to talk me out of it but I pointed out they can run the place without me and could hire temporary help if needed. As you did, I am not demanding my third share. I would certainly like to see where you live and what you do. I purchased a train ticket all the way to Shadowcreek. I'll arrive at 2 p.m. on May 10th. If it is at all possible it surely would be nice to have you there at the station. If you can't be there I'll make inquiries as to where you might be. See you soon.

                              Your brother Alva

Chapter Three - 1907

"Well, I'll be," exclaimed Buck. "It's hard to believe my kid brother is grown up and on his own. It'll be good to see him."

"It's wonderful he's coming. I'll find out all about your skeletons," laughed Edna.

"Not if I don't let the two of you alone," countered Buck.

"The 10th is only a few days away. I'll be glad to help you in any way I can while he's here."

"Thanks for the offer, Eddy. Oh my gosh, I called you Eddy instead of Edna. It just slipped out. You probably don't like it."

"Buck, you make me laugh when you say something just slips out. I love it and wish you would call me that anytime you want to. Would you?"

"I guess so. I'll call you that from now on, except when you upset me and that's never happened yet."

"You rascal," retorted Edna.

* * *

The next day Buck rode to the supervisor's office. Supervisor Bolton handed him some written orders directed to all rangers. Grazing seasons were established for six months, eight months or one year starting May 1st. Payments were to be made before stock entered the range and would be made by bank draft or money order to James Adair, Special Fiscal Agent. Cash could not be accepted. Kent had requested a meeting for all stockmen to be at the grange hall the next day. Buck was apprehensive.

The meeting was called to order by the supervisor with Buck the only ranger present. Kent stated that a Federal court in Montana had ruled it was legal for the Secretary of Agriculture to charge grazing fees on the national forests. The Forest Service could also penalize stockmen for having animals on national forest land without a permit. When he added payments had to be made by bank draft or money order before stock entered the range, a free-for-all almost broke out. Kent could barely control the group and it took Buck all his reasoning and mediation skills to settle everyone down. As the day progressed, Kent and Buck answered dozens of questions. They tried to give good reasons for the "why" inquiries. Buck reminded the ranchers on his district that during the previous year he had met with them individually and had thought everything had been agreed to, even if they hadn't liked it. They admitted that was true. This helped to calm the waters.

An old timer spoke about a non-grazing issue. He understood from some newspaper clippings he possessed that no person could carry a gun or cut a few cords of wood or lumber on the forest without getting a permit from a forest officer. He asked why should we pioneers be barred from privileges that others have? Kent responded that he had a letter from the head forester in Washington, that local settlers were entitled to free use of timber and there was no rule against carrying guns on forest land.

The meeting broke up with a grumbling undertone but at least no one threatened anyone or said they wouldn't obey the rules.

It had been a long day for both forest officers. Kent offered Buck a drink at the Golden Restaurant and Saloon. When they entered the swinging doors, there was Zeke the bartender still selling booze and giving out advice. Two years before he had performed the same function when Kent and Buck first met outside on the street. He remembered the incident and greeted both men warmly, asking them if they wanted the same as before. Buck asked him what he had ordered for a drink at that time. Without hesitating Zeke said it was whiskey.

"Now that's amazing. So, what did Kent here have?"

"Brandy!"

It was now the supervisors turn to be astonished. "Why, that's right. How can you remember that long ago?"

"It's my business. Do you want the same now?"

"Make it a double for both of us," said Kent.

They stood at the bar with one foot on the rail and talked about the day's events.

The next day Buck's return to his district was delayed. The rules of the Service required that supervisor's hold a meeting of all rangers, guards and members of the fire patrol once a year. Kent decided the meeting would be the following day and had phoned the districts earlier. The guards, including Luke, were back on the payroll. It was a convenient time before summer work placed many of the staff out of touch.

They met at the Harmony Grange Hall. There were 11 in the group, representing the four districts. Kent's agenda started with his review of what had happened two days ago at the stockman's meeting. Rangers were brought up to date on fees and permits. The forester had authorized 4,500 cattle and horses and 1,000 sheep and goats on the Maahcooatche Forest for the year.

Rangers were requested to report on predatory animals killed in or near the national forest for the previous calendar year. This was due to a request by the Bureau of Biological Survey.

Another proclamation had been issued that affected the Maahcooatche

Chapter Three - 1907 77

Forest. It added acreage for districts 1 and 2 and a portion of district 4 was transferred to an adjacent national forest. Buck's district was not affected.

Kent was going to conduct a ranger's examination in Harmony. If anyone in attendance knew of a good candidate, the supervisor was to be informed.

Many other topics were discussed. Kent said more timber sales were to be planned and a careful inventory of the timber volume on each district was to be done as soon as possible. He specifically instructed Buck and Hugh Tanner to combine forces for a large timber sale covering parts of both districts 3 and 4. He wanted it sold before winter.

There were many mining claims still to check.

More signs needed installation on trails and boundaries. They were available at the storeroom back of the office.

Near the end of his narration, Kent said that Mr. Pinchot regarded boundary survey work as one of the highest priorities. All the rangers needed to get on with their own district boundary surveys and have them completed by the earliest possible date.

Finally, the supervisor said, "Well, men, I've saved the best 'til last. I want to thank you all for performing well and as a reward each ranger is to receive an additional $25 a month—which means your yearly pay is now $1,200."

There was a yell of approval from the four rangers. Kent assured the others that more pay was in the works. The meeting was adjourned. Buck and Hugh got their heads together for their timber sale and worked late that evening.

The next day Luke and Buck rode back to Lakefield station. They were determined to get the telephone line repaired. Luke had no experience using snowshoes and even though he was strong and could walk many miles in a day with a full pack didn't mean he was prepared for a snow trek. The muscles required were not like regular hiking. It took a day to get acclimated to walking with large shoes. The three animals carried food and supplies. The snow on the ground was melting rapidly.

The first day they repaired several breaks on the line from fallen trees and the telephone still didn't work. The second day they left the animals in the barn with plenty of food and water and tried to get as far as they could, camp overnight and hike back to the station the next day. They found the going slow with loaded back packs and discovered wire breaks which were repaired using a Western Union splice. That night a snag was felled and a warming fire was started. During the next few hours, either Buck or Luke kept pushing the pole toward the heat of the fire. They stayed comparatively warm until morning.

About halfway to Crescent River they figured it best to come in from the other end and finish any repairs. As they retraced their steps, the snow got deeper. That night more had fallen in the higher elevations.

Luke was about 50 feet behind Buck. He thought something was moving off to his right so he turned to check it out, went about 100 feet and disappeared. Buck heard a short cry, turned and didn't see his partner. Quickly backtracking, he followed Luke's imprints to a point where they ended in a deep depression. Snow had completely covered any sign of the guard. Luke had a problem. The heavy wet snow had not only covered him but prevented his movement. His arms were raised as he dropped in the crevice and the snow from the sides cascaded on top of his prone body. Since his arms were now by his head he did manage to remove the snow from his eyes, mouth and nose.

Buck could see where the snow covered the tracks and knew Luke was under it. Carefully stepping down into the crater he quickly pawed several inches to the side. In a couple of minutes the tip of a snowshoe appeared. Assuming it was still attached to Luke he figured where the guard's head would be. With renewed energy he started digging. The snow flew and there was Luke staring up from his white coffin.

"What took you so long?" he asked, half jokingly.

"You can't hide from me like that, you old goat," answered Buck.

"I can't move, so quit talking and start digging you, you, you!" Luke couldn't think of any name to call his boss so he stopped talking.

It took five minutes to get him out and another ten to rest and recover. The two arrived at Lakefield in the early evening. The telephone crank showed some resistance and finally there was an answer from the exchange. They let out a whoop and told the operator she sounded wonderful.

The following day was May 10th, so the ranger and guard rode back to Shadowcreek in the morning. Buck showed Luke the plans for the six fire cache structures and asked him to buy the material and tools. He wanted them placed throughout the district that summer.

The 2 p.m. train was 15 minutes late when it pulled into Shadowcreek station. Buck and Edna were there with Titus and Stub. Two men and a woman disembarked. No Alva. Buck jumped up the steps between the only two passenger cars. He looked to his left and then to his right. He thought he saw Alva slumped in his seat near the rear. Signaling Edna to head that way from the outside, he strode down the aisle past three or four passengers. There was Alva, bleary eyed and trying to talk. Buck

Chapter Three - 1907

handed Edna the luggage and half carried Alva down the stairs to the platform.

The doctor's office was a block away so Buck carried his brother and Edna led the animals. Doc Cleary was an older gentleman who had practiced in town for almost 40 years. He was loved and admired by the citizens. In a few minutes he determined Alva had been drugged. Giving him something to swallow the visitor started to come around quickly.

He told Buck the following story: Two men boarded the train at Harmony. They sat next to him and started talking. Bringing out a flask of whiskey they offered him some. The more he refused the more insistent they were about drinking a round together for friendship. To get rid of them he finally took a swallow. That's the last he remembered until he realized Buck was dragging him off the train.

Alva felt in his pocket. His money was gone. From Alva's description of the two men, Buck remembered they were the ones who had departed the train. He figured they were walking back to Harmony, so he paid the Doc and asked Edna to take his brother and mule to her house until he returned. With his master on his back, Titus took off like a thoroughbred.

The two men were about three miles from Fish Cut when Buck spotted them on the road ahead. The large one, who Alva had identified as the leader, was on the left. He slowed Titus to an easy gait and withdrew his rifle from its sheath. As he approached, the men looked around and parted, one to each side of the road. It was a perfect move for Buck. Taking the barrel of the rifle in both hands he swung the butt around and smashed the large outlaw on the head. Down he went like a lead sack. Before the second outlaw knew what was happening, Buck jumped off Titus and tackled him. He started yelling and pleading as he was dragged over to his partner. The ranger removed his handcuffs and snapped one side to the right wrist of the unconscious one and the other to the left wrist of the vocal one. Next he felt through the pockets of the first man and found a roll of money. The second man gave up another wad and Buck stuffed them both into his saddlebag. He hadn't asked Alva how much money was missing but he sure wasn't going to give any back to the outlaws. He also removed the flask of whiskey from the large man's shirt. Luckily it hadn't broken.

After the two men regained their composure, Buck ordered them to walk in front of him and head for Harmony. It took more than two hours to deliver them to the jail and explain to the deputy what had happened. He knew Buck well by this time, so took him for his word. Handing the deputy the flask, Buck explained it probably contained drugs and could be used for evidence.

By the time the horse was sheltered at Ike's Livery Stable, it was dark. Edna had previously dropped off Stub. Alva was his old self when Buck arrived at Edna's and the three of them talked for another couple of hours with food furnished by the hostess.

When Alva protested that it was more money than he had lost, Buck just winked and told him he must be mistaken. The two outlaws had given up the loot easily so it must be all his.

The brothers roomed at the Pilot Hotel and postponed talk of the future until the next day. Before falling to sleep, Alva thanked his brother for not scolding him for being such a naïve fool. Buck's only comment was that every experience in life should make one a little smarter.

* * *

Alva wanted to stay and tag along with Buck as he worked. There was no money to pay him and there was a prohibition against nepotism. He was signed as an official volunteer. But first he needed a horse and saddle. Buck remembered a really nice palomino that he had used. Elroy just might be willing to sell him. After some friendly bartering, the deal was made and Alva was the owner of Lightning. He had brought his own rifle from home and with the purchase of a secondhand saddle, little brother was in business.

Luke had built one fire cache structure in Shadowcreek. It was functional but too large for an animal to carry. The decision was to dismantle the cache and take the material and tools and construct them at each site. Eventually, this was done but it took all summer and fall to complete the job.

The work priorities were discussed. At the forest meeting several days before, Buck and Luke gathered a stack of boundary posters. Buck had also been given a list of seven more homestead claims to survey and process but his main task was to get the large timber sale started. It was agreed that as Luke patrolled for fires, he would place signs at strategic locations, would map and sketch portions of the district for agriculture land possibilities and make observations on timber volume without actually cruising the whole areas he happened to be in. These acres he would identify on a district map so that neither man would overlap work.

In the meantime Buck contacted Ranger Hugh Tanner. They agreed to meet at Barnesville.

Buck, Alva, Titus, Stub, Lightning and an extra pack horse left for

Lakefield with a full load of supplies and food to last a couple of weeks. It was evening when they arrived at Lakefield and Alva was given a quick tour. He marveled at his big brother's many talents.

It was a long trip to Barnesville, so an early start was made. The trail was in generally good shape after the previous year's maintenance, although a series of late spring storms during May had created problems as the trail gained elevation. The animals were slipping on ice and snow.

At one point the trail and a stream came together. On one side the trail butted up against a high cliff. On the other side it dropped into the streambed twenty feet below. Solid ice covered the tread. There was no way the animals could safely walk up the stream, so Buck told his brother they were going to tight line the horses and mules for about 60 feet until the cliff opened up again.

Tying a rope to Titus's neck and carrying it across to safety, he instructed Alva to tie another rope around the horse's tail and wrap the other end a couple of times around a tree.

The ranger took the lead rope and did the same. As Alva let out a few inches at a time, Buck pulled on his end and took up the slack so that Titus was always on a tight line. Since each animal had to be inched across the ice trail the same way, it took a couple of hours. The loads scraped the cliff, but with patience and calmness, the feat was accomplished with no accident or injury.

The Barnesville Station was owned by Hugh. He had continued his ranching after going to work for the old General Land Office. After the land was transferred to the Department of Agriculture, he curtailed his ranching activities in favor of continuing his ranger duties. They reached Barnesville at dusk. The six travelers were tired, thirsty and hungry. Hugh greeted their arrival and directed Buck to the barn. After the animals were cared for, the brothers ate a hearty meal at Hugh's invitation at his abode. Mrs. Susan Tanner was a pleasant, nice looking woman with a fair complexion and silky, dark hair. She was glad to meet the ranger next door she had heard so much about. Buck didn't say a word but assumed she was referring to the husband's kidnapping at Lakefield to help construct the cabin roof. The next morning Hugh introduced Guard Tom McCarthy to Buck and Alva. The four of them were to cruise a timber sale covering portions of both districts. It would be the largest timber sale ever offered on the Maahooatche, about 5,000,000 board feet of ponderosa pine, sugar pine and Douglas fir. It took about a month to complete the cruise. Information as to species, diameter and number of 16 foot logs in each tree were gathered in the field. Volumes were calculated later in the office.

The entire prospectus was published in several local newspapers and was distributed to Western timber operators. Three acceptable bids were received and the Split Rock Timber Sale was awarded by the forester to the Reliance Lumber Company headquartered in Ingot.

To gain knowledge for future large sales, all four district rangers and the supervisor attended the first meeting with the company representatives, President Collier Newton and two other employees. Kent ran the show and everything went smoothly. Because of logistics, it was agreed that the Barnesville District would be responsible for administering the sale. This included marking, scaling and inspecting. The supervisor stated the contract called for the disposal of brush. He emphasized the handling of brush should not fall behind the cutting and removal of timber. The disposal may vary, depending on the area. It may be found advisable to lop and scatter the tops to prevent erosion or to favor reproduction. Burning brush piles was required. Kent continued his speech by reminding the contractor that he is required to use the method of logging that does not damage the reproduction or is likely to cause erosion. He ended by saying that Hugh Tanner would explain the marking of trees and scaling.

Hugh stood up and said, "I want to thank you all for your interest in our big sale. I expect three knowledgeable employees from the Washington office to arrive within a day or two. They will administer the sale, along with Guard Tim McCarthy, if he is available. We will be marking timber ahead of your fallers, Mr. Newton. Our marker will use a hatchet with a US die etched into the metal at the hammer head end. He will blaze a tree to be cut at eye level and stamp US on the butt or bottom log of the tree. Then he will blaze the same tree below the cut line on the stump and again use the hatchet for another US brand. We don't expect any trees to be cut that don't show a US. We'll be checking stumps after you fall the tree. Scaling will take place on the railroad flat cars. We will use a 20 foot ladder. The end of each log will also be stamped with a US to show it has been measured to estimate the amount of lumber it contains. Does anyone have a question?"

Since no questions were asked, Hugh requested Collier Newton explain what logging methods would be used. The contract did not require a particular method.

Collier stood and thanked the Forest Service folks for their hospitality and detailed explanations. He continued, "Since we plan to move all the logs to the railroad landing, it will be necessary to use at least three methods of transportation. Before starting any specific area we will meet with the Forest Service inspectors and agree on which method is best. We all know it is impossible to remove large logs from the forest without doing some damage to smaller trees and vegetation. However, we intend to keep it to a minimum as the contract specifies. The first method will involve the big wheels. These must be used in relatively flat areas since the stringer tongue type that we use does not have a braking system. These big wheels with a diameter of from 10 to 12 feet will be pulled by four horses. The second method we will use involves oxen. Our bull teams will haul logs from the woods to the railroad landing over skidroads. Teams may be six, eight or ten bulls depending on the load. We don't use a harness. A yoke fits across the necks of each pair of bulls. This is hooked to a heavy chain that in turn is hooked to a big log or even more than one. The third method is by steam donkey. They have the capability of operating with more than 1,000 feet of cable. Does anyone have any questions?"

Buck had never seen a steam donkey hauling logs from the forest. He asked, "Mr. Newton, would you explain the steps in how a log is pulled from the woods to a landing by your last method?"

"I'd be happy to," replied Collier. "Our skidding takes place when the cable gets to a bucked log and a man called the hook tender jumps at the heavy block and pulls slack into the line and drops a big hook over the choker or cable. Our choker setter has previously looped the choker around the log. The hook tender will then raise his hand and jump clear. The whistle punk, or boy who mans a whistle wire running back to the donkey, will give it a jerk. At that, the donkey puncher will pull the lead line taut and let in the steam. The engine should then make a great deal of noise and start rocking on its skids as the line is yanking the log along the ground. With the cable winding around the donkey's drum, the log will eventually appear and hit the landing. Any other questions?"

There were none.

Kent thanked everyone for being there and declared the meeting over. Afterwards the supervisor suggested that all four rangers, plus himself, go for a campout away from the station. There they could discuss problems and socialize. With a twinkle in his eye he said the emphasis would be on the latter so everyone should bring their own libations and food.

Barnesville was a sleepy town of about 500 souls. The ranger station was a half mile from any store, so the visitors walked to town, loaded up with food for supper and breakfast and enough liquid to wash it all down plus more for an evening of fun and revelry.

The group consisted of Kent, Buck, Hugh, Ralph, Tim and Alva. Hugh led them to a beautiful meadow split by a sparkling stream and surrounded by big timber. Evening was approaching and a campfire built. There were no teetotalers among the group. Kent discussed official information for about ten minutes, then said tale time was the order of business for the rest of the evening. Tim and Alva were eager with anticipation. They had heard there was nothing compared to a group of rangers trying to top one another with wild and daring tales. It was an entertainment privilege that few people would ever experience but would never be forgotten.

As the host ranger, Hugh began, "Well, you skeptics may not believe this but I was told this story by a very reliable friend who swore to its authenticity. It seems that a fellow ranger working all day in the woods decided to spend the night sleeping on the ground. In his dreams he imagined a humming type of machinery and then he got a whiff of a horrible putrid stench. On opening his eyes there was a large cougar looking directly at him with his nose about six inches away straddling the ranger's body. The cat was purring and glad to have company but his breath was indescribable. The ranger reached under the pillow for his revolver and accidentally pulled the trigger. Feathers were flying as did the cougar, throwing large quantities of twigs, cones, dirt and needles over the bed as he bounded away." Everyone guffawed and snorted a shot of booze.

It was Kent's turn next. He pulled out an envelope and said the enclosed letter had been received a few days ago and he guessed the group would appreciate the contents. It was dated June 26, 1907, and addressed to the Right Honorable Kent Bolton, ESQ. Lord High Protector of the Wilderness and Chancellor of its Resources.

Most Estimable Sir:
Some time ago I addressed to you a letter requesting the privilege of Procuring wood from the Forest Reserve for fuel.
I came into this region recently, and have not fully learned the customs of the people or the land; but I am informed that I have not the privilege to cut a tooth pick from a piece of decaying drift wood without a special permit from your Royal Highness.
Now I do not quarrel with your customs, tho they be diverse from all peoples, but I most sincerely desire fuel.

I know of a tree, both high and ancient, which has ceased to respond to the caresses of the rain and the sunshine; leaves are withered and falling to the ground; whose branches are bare and distorted; whose roots are loose in their earthen sockets, and whose lofty trunk sways most pathetically in the chill, congealing breezes.

That this venerable pile may not fall into an unhonored decay, but may be respectably cremated, and incidentally, my family may have the necessary means for cooking their simple meals and warming their humble apartments, I respectfully but earnestly petition Your Excellency for the privilege of removing the aforementioned timber from its present site, which is on a slightly elevated point of land six miles southwest of Harmony, to the land adjoining my present residence, where it will be cut into convenient form, and treated as above indicated.

If I may not have this tree, or some other tree, or at least some broken branch from off some solitary stem which is doomed to swift and certain decay, please advise me where and how and when I may procure an empty box, a broken chair or a bit of brittle straw, lest my family and myself come to grief thru lack of fuel.

I enclose herewith a stamped envelope addressed to myself, and a sheet of paper where on you may ascribe the magic words which will at least enlighten me.

> I am, Dear Sir,
> Your Obedient Servant and Loyal Subject,
> Lloyd Garrison Knight

When Kent finished, the rangers laughed so hard they hurt. The group agreed it was the most sarcastic piece of wit they had ever heard. A masterpiece said one.

Ralph was next to tell his tale. "It seems that Old Joe, a fire guard, was working all summer out of a tent. He would work all day on the assigned drift fence construction and ride back to camp in the evening. Beans and more beans was his diet. He was getting sick of beans and sick of being lonely. One day he rode to his house in town and persuaded Mrs. Joe to come up and stay with him the rest of the summer. They could both sleep in the tent and Joe could have some good home cooked meals and even get some dirty clothes cleaned. He was in guard heaven.

One night a bear came prowling around, banging this and that and generally keeping the inhabitants awake most of the night. Joe didn't like using his rifle, so he obtained a 2x4 board and laid it by his bed. The next night Joe was sleeping peacefully but Mrs. Joe had to heed the call of nature. Being timid, she stayed by the tent, went around the corner and squatted. Joe heard something outside, didn't notice his wife was missing but did notice a bulge in the canvas and some movement. It was now or never he said to himself. Picking up the 2x4 he swung it with all his might, with a resulting thud and then quiet. For the rest of the summer Joe was lonely, ate beans and worked in soiled clothing."

Everyone laughed and took another swig of John Barleycorn.

It was Buck's turn. The others were anticipating a good story because Buck was the most animated and could talk with drawls and accents.

"This is a change of pace, guys. I'm not going to tell a story about the Forest Service. Anyway, I've had more experiences on the range and at ranches and farms. It seems this big city fancy lawyer was duck hunting in rural Oklahoma. My friend, farmer Jack Johnson, told me this story. I know it is true since Jack was the farmer involved. Jack had heard some shooting and ran out to see a bird drop from the sky and land on his land. He noticed this dude with a shotgun on the other side of the boundary fence, about to climb over and retrieve the dead duck. He yelled, 'This is my property and you better not climb over that fence.' The exasperated lawyer replied, 'I don't care what you say, that's my duck and I'm going to get it. If you don't let me, I'll sue you for every cent you have.' Jack smiled and said, 'We have a way to settle disputes here in the Oklahoma Territory. It's called the Oklahoma three kick rule.' The lawyer asked, 'What's the Oklahoma three kick rule?' 'Well,' farmer Johnson went on. 'It goes like this. First I kick you three times, then you kick me three times, and so on until someone gives up.' The lawyer thought about it for a moment and, studying the old farmer, thought he could easily beat him, so he says, 'OK, you're on. I'll abide by the rules.' Jack told the lawyer to come on over to the fence and he would start. His first kick planted the toe of his heavy work boot into the lawyer's groin and dropped him to his knees. His second kick nearly flattened his face and the lawyer fell prone on his stomach. The farmer then kicked as hard as he could directly into the kidney. This almost caused the lawyer to give up but he summoned every bit of his energy and managed to stagger to his feet. 'OK, you old coot, it's my turn.' Farmer Johnson smiled and said, 'Naw, I give up. You can have the duck.'"

The audience of five roared with laughter. The way Buck had thrown

Chapter Three - 1907

his body around, telling the story, was as funny as the tale itself. Another swig was called for.

The first three participants wanted both Tim and Alva to tell a story. Buck didn't say a word. He figured his friend and brother were enjoying themselves listening and didn't want to embarrass them. After several words back and forth Tim finally relented. He told the story of the first time he met Buck and what had happened at the timber trespass case last year. Although he had warned the others it wasn't a funny story, he told it well and was given a round of applause when finished.

Alva tried to avoid telling a story but eventually relented. His wasn't funny either, he said, but thought the group would think it interesting. Then he told his experience on the train, being drugged, robbed and his brother Buck catching the criminals and getting his money back. There was another round of applause.

No one wanted to turn in for the night. The booze was still flowing and the fire was still hot. Continuing in the same order a second round of stories was told. Three were funny, one was serious, one was a mystery and one was a tragedy.

It was now quite late and Kent suggested everyone turn in. No one was going to contradict the supervisor, so with one more log on the fire and a quick trip to relieve themselves, the six inebriated tale spinners spread out around the circumference of the glow and were soon off to la-la land.

* * *

The Washington office instructed that all rangers were to use the new 874 ranger's notebook. It contained several report forms, free use permits and scale record sheets. It was to include tables of local prices for timber products such as poles and shakes. Fire reports were also to be made on the 874 notebook forms. More emphasis was placed on complete reports for all fires. No clear definition of a statistical report had yet been established, so all false alarms were reported as fires, as were fires that were out before found. Causes of fires had not yet been formally classified. Therefore, for review purposes, breaking man-caused into incendiary and other categories gave the only significant figures. One report submitted listed the cause as "a man and a match." Specific directions were issued that the 874 time report was not to replace the ranger diaries.

Buck tried to keep his daily diary on a daily basis. This was impossible when there were fires or other emergencies. He did not write flowery descriptions, as his story telling was prone to do. He was succinct and to the point. There was never any question as to what he meant or what he actually did. He didn't think it was really necessary but considered it a part of the job and didn't object like some.

Alva had been following Buck around for four months. Buck watched his brother closely to see if he liked the work, didn't complain and was mature enough to do the job of a fire guard. He thought Alva would make an excellent Forest Service employee but he wasn't going to urge him to make any decision.

One day in August, Buck, Luke and Alva were at Lakefield eating the evening meal together for the first time in a couple of months. Luke had been doing a variety of tasks other than fire patrol. He had erected and stocked three of the six fire caches around the district. Many of the trail and boundary signs had been installed. He had mapped almost a third of the district for compliance with the Homestead Act and had estimated volumes of timber as he went by running lines through 640 acre sections.

Buck had to make some priorities. He and Alva had also drawn sketches for agricultural areas that would probably pass under the Act and estimated timber volumes in many sections. Adding up the work he had done with Luke's showed there was no way before winter that both jobs could be completed. He and Luke decided to concentrate on the Homestead Act requirements and put the timber volume analysis aside until the following year.

A few months before, he had vowed to do it all in 1907. He hadn't realized the number of days it would take with just two people, nor had he realized there would be so many interruptions. He was not discouraged, however, just more realistic. The other districts had assistant rangers and at least two guards. He finally accepted the fact that he needed help.

That evening at the station Alva told the other two that he would like to work for the Forest Service in some capacity. He had been thinking about it for some time but wasn't sure he could do it or that his older brother would agree. Buck told him he would make a good employee. He had watched how he handled horses and mules, how he worked long hours, didn't complain and was a fast learner. Alva knew he couldn't be a ranger until passing the test. He could be a fire guard, however, and asked Buck if he could put in a good word at the Harmony office. Buck had an idea for the rest of the year but needed to get permission from Kent or Garth.

Chapter Three - 1907

The next morning he called the supervisor's office to make sure one of the supervisors would be there. Lucy answered and said Garth was gone. He had been transferred to a forest supervisor job on a northern forest. Buck passed the word to Luke in a whisper as Kent answered the telephone. The ranger asked if he would be in the office the next day; he had several important things to discuss. Kent said to come ahead. Buck packed and gathered all the analysis, reports, diaries and completed forms. He instructed Luke and Alva to keep patrolling for fire and to keep mapping for agricultural land on the district. Buck left early to see Edna and get to Harmony by evening.

The main trail from Lakefield station to Shadowcreek was in good shape. The public was using it more and more for recreational purposes. About a mile from the district boundary, before the trail converted to the road, a little stream named Blackberry Creek ran through a meadow. It was a tributary of Shadow Creek and had hundreds of blackberry bushes along its bank, running for hundreds of feet. The townspeople would hike or ride with pails and baskets for berry picking. Edna had made Buck a blackberry pie more than once. He had some of her jam at the station.

As Buck rode out of the timber into the meadow, about 100 yards from the creek, a wild looking, desperate, dirty middle aged man jumped out from behind a tree pointing a rifle directly at the ranger.

"Get off that horse right now," said the man.

"Why should I?" asked Buck. "Stop pointing that rifle, it might go off."

"Dang right it might go off this very second if you don't get down off that horse."

"If you want my money, here you can have it. It's not much but it's all yours."

"I don't want your money, I want your horse. That's a mighty fine looking animal so get down NOW!"

Buck could hear the desperation in his voice. If he wanted Titus, the stranger was in for a big surprise. He jumped off and moved away from the horse as the outlaw swung his rifle around like a man possessed and kept shouting.

As soon as he sat in the saddle and grabbed the reins, Buck whistled. Titus took off like a shot. The man lost his rifle and was holding on for dear life, yelling and pleading for the next hundred yards. The trail turned sharply to the right just as it passed the first blackberry bushes.

At that moment Buck whistled again and Titus stopped immediately but the rider didn't. He catapulted off the saddle like a cannon ball and landed in the middle of the bushes about ten feet from Titus. It was also at the point where the water passed through the patch. Buck whistled again and Titus trotted back.

The ranger felt like leaving him there in the bushes. His compassion ruled his sense of justice. Buck dismounted to survey the damage. The outlaw was a mess. Thorns had sliced him from head to toe. Much of his clothing had been ripped. He was bleeding profusely and said he thought his leg was broken. Buck lassoed him and Titus dragged him out onto the trail. Mud was mixed with blood. His leg wasn't broken but two fingers were, plus his nose. He looked as if he had lost a fight with some barbed wire and a bear.

The man said his name was Herman Woodford and that he needed transportation to get back home about 200 miles away. He knew folks used the trail so he had waited for the first person to show up with a horse. He figured he was in luck after Buck showed up first. The horse was a beauty. It would be easy.

He begged Buck to let him ride Titus because he was in so much pain. Buck said ordinarily he might consider it but not with Titus. Herman would have to walk the three miles to town. It took them an hour and one-half.

Buck went directly to Doc Cleary's office. Since the doctor was out delivering a baby they headed for the Reverend Eugene Findley's home. He knew Mrs. Findley would take some pity on the injured outlaw and care for his wounds. The sheriff wasn't too happy to hear that Buck had another prisoner but said they would send a man over that evening to the Reverend's house with an extra horse to bring Mr. Woodford back to jail in Harmony. Buck left the man's rifle for the sheriff, thanked the couple for taking care of his problem and left to find Edna.

It was a happy reunion. They hadn't seen each other for several weeks and had a lot of catching up to do. After dinner, Buck excused himself. He had to stay in Harmony that night so he could talk with the boss first thing in the morning. He promised to stop in Shadowcreek on his return to Lakefield District.

He was waiting the next morning when Lucy arrived to open the office. Before Kent appeared he asked her if he could practice his typing skills. It had been awhile but his fingers were as nimble as ever. He was concentrating so hard that Kent entered unnoticed from the rear door.

Kent stood behind Buck for a few seconds and then said, "Young

Chapter Three - 1907

man, we have a real good clerk; but if ever we need another one you're hired."

Buck jumped in surprise and replied, "Darn it Kent, don't sneak up on me that way. As for being a clerk, I wouldn't mind but I'll be danged if I'd correct all your spelling and grammatical errors. I've seen some of your rough drafts."

"Now don't get smart young man, you're not hired yet and as for my writing, let's ask Lucy for her opinion."

"Don't get me into the middle of your ridiculous sniping," said Lucy. "You two act like a couple of kids."

Buck knew Kent's writing was almost perfect. He also knew that he could push his boss's buttons and not be scolded, if it was done in fun. He admired Kent for his easy manner but he also respected him for his toughness when necessary. He thought Kent was the best boss he had ever worked for. On the other hand, Kent thought Buck was the best ranger he had ever known. Both men kept their thoughts to themselves.

There were many things to talk about. Kent told Buck that Garth had been offered a forest supervisor job up near the Canadian border. It was an opportunity he couldn't refuse. It had happened so fast, there wasn't time to notify the other rangers except to leave telephone messages. When Buck asked if there was a replacement, Kent replied he didn't know but that he would be able to do it all for a few months.

The supervisor also said he had heard by the grapevine there would be six district[13] offices opening around the country next year. This would mean an extra layer of command so that all forest level correspondence and decision making would be between the district headquarters and the forest. It had become unwieldy for all forests to report directly to the Washington office. At that time, he didn't know what city the Maahcooatche would report to. Buck also received a folder full of Homestead Act claims. Kent mentioned that the state was going to issue deer hunting licenses for the first time that fall. They would be available for $1 from the county clerk. The department of fish and game had asked all forest rangers to help in checking compliance. Both men looked at each other and almost at the same time said, "We'll just put it on the list as one more duty."

Kent added he had one more thing for Buck to check. He had heard there was a potential conflict between the cattlemen on the eastside allotments and someone coming in to graze sheep in the same area. Buck perked up at this statement. To date he had had no sheep on his district and had only read about other folks' problems on the subject.

"Can you give me any specifics?" Buck asked.

"Not yet. I've only heard a rumor from a friend of mine but I believe it's more than just a piece of gossip. We can't afford to ignore it, Buck. I want you to stop everything and give this first priority.

"The current *Use Book* states, 'National Forests in which grazing is allowed will be divided into districts approved by the Forester, who will determine the kind of stock to be grazed in each district. The supervisor will make such range divisions among applicants for the grazing permits as appear most equitable and for the best interest of the National Forest and its users.'"

"Where is this mixing of sheep and cattle supposed to be?"

"The northeast corner of your district."

"That's probably Poker Flat. Jug Handley and others have been using it for cattle. He only has a few head but he's not one to be pushed around. Before I leave though, there are a couple of things I would like an answer to. As you know, my brother has been riding with me for the past four months. He really helped and has done an outstanding job. You know I wouldn't say that if it weren't true. Anyway, he has expressed a desire to join the Forest Service in some capacity. He wants to take the ranger exam when the next one is held but in the meantime he would like to be a paid guard. Since I'm short on personnel, is there a chance he could be placed on the Lakefield District, even though he's my brother?"

"Buck, I know your personnel problems but we just don't have the funding, even if I could get a waiver on your relationship. I would suggest Alva go to as many fires around the forest this fall as he can. That way he can get paid from fire funds. He's already been on a few I know, but he can make himself available to the other districts and even other forests. What about it?"

"Thanks, Kent. I'll tell him. Will you put him on the list for the next exam?"

"Sure will. With you training him, he should pass easily."

"I've turned in my diaries, Luke's and my sketches and drawings relating to the Homestead Act and some timber volumes on quite a few sections. I know we won't be done this year but I'll try my darndest for next."

"Buck, I suspect you're spending too much time trying to be super accurate on your timber volume report. I believe the Washington office wants us to do more estimating than you're doing. Take an area, make the totals, then take a like area and estimate. Don't run lines on each and

Chapter Three - 1907

every section of your district. It would take several years for the two of you, with all the other work you have."

"If that's all right, I'll do it."

"I've made that decision on what I think our priorities should be. There will be more and more added tasks piled on us every year from now on. I don't foresee funds for the necessary manpower to do all the work. For example, you still need to check on mining claims. We've let them ride on this forest in place of other work but we've come to a point we can't ignore them any more."

"Sounds as if we need more days in the year."

"Yea, but I have faith in the four of you rangers. We'll get there."

"You can count on me, Kent. I know we can do it. I've never worried about the number of hours or the number of days we work and I'm not going to start now. It's still morning and I want to get to Jug Handley's house by evening."

"Thanks, Buck. Keep in touch."

Buck said goodbye to Lucy and grabbed a sandwich to take with him from the Golden Restaurant and Saloon. The ranger, Titus and Stub were on their way to Shadowcreek for a quick goodbye to Edna and then north to the grazing lands.

The day was sunny, the breeze was refreshing and the stillness was broken only by the steady hoofs of horse and mule striking the dirt road. It was mid afternoon when they arrived at Jug Handley's spread.

Buck had never had any trouble with Jug. They trusted each other. The first year, Buck had counted animals on most of the districts range allotments. The number Jug had reported and what he actually had were the same. He hadn't counted any since.

The dogs at the ranch gave Buck away. Jug stepped out of the front door and, when he saw who it was, created a big smile and welcomed the ranger. He was in his thirties. His hair was receding prematurely. He was large boned at 5 feet 11 inches. His talk was slow but his walk was fast. Buck and Jug had hit it off right from the beginning. The family was just sitting down to eat and the ranger was invited to join in. Stories and small talk were the conversation at supper. Jug's family consisted of his wife and a well-mannered son and daughter. They all had red hair.

After supper, Jug and Buck went outside to take care of the animals. Jug wanted him to stay the night in the house but Buck would have none of it. The barn would suit him just fine. When their chores were done, Buck got serious and asked about the sheep.

Chapter Three - 1907

"Jug, I heard from my boss that there are some sheep in the same district you're grazing. I really don't know any facts. That's why I'm here."

"I thought that was why you were here. Yep, there is a new rancher with a band of sheep in the area and he just drove them on government land. I don't want my cattle mixing with those sheep and I guess if you can't fix the problem, a bunch of us ranchers will go to his spread and run him out even if we have to shoot the sheep."

"Promise me you won't do anything until I've been given the chance to resolve the situation."

"All right, I promise, but don't take too long, Buck. We've got a few weeks to go before they come off the range."

"What's this guy's name and where does he live?"

"His name is Leo Pitman and he lives at the old Dodge Ranch. You know, where the road from Shadowcreek meets the road going west to Barnesville."

"Well, I'm going to get a real early start so you won't see me in the morning, Jug. Thanks so very much for the supper and your hospitality."

"Let me know what happens, Buck. Good luck."

As Buck rode north the next morning, he thought about what he hoped to do. First thing was to check the facts from the sheepherder's side. He had read about range wars, when sheep were shot by cattlemen and cattle were shot by sheepmen. Even men had been shot. He needed to defuse a potential volatile situation. Leo Pitman's sheep ranch was a motley looking place. The buildings were in need of repair and the place had a bad smell. Buck looked around carefully before dismounting. There was no one moving that he could see. Knocking on the door, he heard someone stirring about.

"Hello, I'm the ranger on this district and would like to talk to someone."

No answer.

"Hello. I know there's someone in there and I won't go away 'til you come to the door."

More movement and then the door opened slowly. A young woman looking 50 said, "If you're looking for my husband, he's not here."

"Where might I find him," asked Buck.

"He's probably out with his sheep."

"If he comes back before I find him, tell him I'm looking for him. Name is Buck Stonewall."

"I will."

Buck left.

The early grazing divisions or districts on the Maahcooatche were made in the supervisor's office by representatives of the stockmen and the forest officers. Maps of all the districts were set up and descriptions of each one were written up. The rangers had to check these maps in the field and Buck had done so the previous year. When problems arose, the ranger was supposed to meet with the folks concerned and settle the difference right on the ground. If the two sides couldn't agree, then the ranger settled it in their presence and made the change, if any, on the map.

Buck headed straight for the Poker Flat district. The first animals he saw were sheep. He yelled Mr. Pitman's name. No answer. He yelled again and heard something behind him. Turning around he was looking into the end of a double barrel shotgun. He kept calm.

"Are you Mr. Leo Pitman?" asked Buck.

"What if I am."

"Well, my name is Buck Stonewall, ranger of this district and I want to talk to you about your sheep."

"What about them?"

"Well, Mr. Pitman, I don't want to talk while you're pointing that gun at me, so please put it down."

"I'll put it down but it's going to stay in my hands."

"Why don't we dismount and talk under that tree?" Buck pointed to a large oak about 40 feet away. "It will give the animals some rest."

"All right, but no funny stuff."

Leo brought his shotgun with him. Buck left his rifle in its sheath. Leo Pitman was a rather short, gaunt looking man in need of a bath. He never smiled nor seemed to enjoy anything about his lot in life. He looked tired and troubled.

"I've brought a map of this portion of the district we are in right now," said Buck. "I notice you have some sheep grazing here and was wondering if you have a permit or have paid any fee to the government?"

"Listen, Mr. Ranger, I just bought my place. Don't know nothing about any permits or fees. My sheep have to eat somewhere and this is open public land, ain't it?"

"It's public land, but it's not open to just anyone with any type or number of animals, without getting a permit and paying a fee. This particular district is set aside for cattle only. There are three stockmen using the area for their cattle. One of them made the complaint about

Chapter Three - 1907

your sheep moving in. This area has reached its maximum number of cattle for forage, so if I allow any more animals it will become overgrazed. The results would mean a ruined range and no one could use it for awhile. You will need to make an application for a grazing permit and when it gets approved you will have to pay a fee."

Mr. Pitman listened as Buck spoke but was getting madder all the time. He said, "I'm not going to move my sheep or pay a fee or get a permit or any of that government jargon. So what are you going to do about it?"

"First of all, I'm going to issue you a citation. Next, I'm going to count your sheep as best I can and charge you damages for trespassing on national forest land."

Leo didn't change expressions, so Buck decided to call his bluff. "Next I'm going to the sheriff's office and bring back a posse and you will be placed under arrest and your sheep will be confiscated. When I leave here I plan to immediately go back to your house and inform Mrs. Pitman that she should start to pack because the government will be taking your house and land for collateral." Buck knew he was taking a calculated risk. He figured Mr. Pitman didn't know the law and might even believe him.

The last two statements got results. Leo put his shotgun down and said, "Wait a minute, Mr. Ranger. There has to be a better way to solve this."

"Yes, there is. If you do what I ask and are honest and don't give me any more crap, then we can solve this, but it will still cost you some extra money."

"What do I do?"

"First tell me how many sheep you have and when you placed them there. Next I want you to remove them immediately back to your spread. You will have to pay a fee for the time you grazed them on government land. I believe there is an area north of here called the Nelson Camp district that is void of any animals at this time. We decided to give it a rest to let the forage recover. It's a little farther from your place than here but it can be used for sheep as well as cattle. I'll check it out and if I determine it can take your sheep, then I'll let you know. What about it?"

Leo thought before speaking. "I don't have enough feed at my place."

"I'll start working on it right away but I want you to start moving your animals right now. Are you willing to seal our agreement with a handshake?"

Leo hesitated and then held out his hand. The deal was sealed and Buck relaxed.

As Buck walked back to Titus he turned and asked, "By the way Leo, would you have really shot me?"

"Probably not, Mr. Ranger. It would have only brought more of your kind and I don't have enough shells." With that, Mr. Pitman doubled over in laughter.

Buck thought to himself that Leo wouldn't be laughing if he knew what he had said was only a bluff. He also thought about his backup plan. At this time he said out loud, "Thank God!"

After some quick sheep counting the ranger headed north to the Nelson Camp District. It was still morning and he spent several hours checking the ground cover and natural barriers. It could be used for Mr. Pitman's sheep.

Buck made camp for the night, next to the east bank of Shadow Creek. The water level had dropped considerably since spring but he still had at the back of his mind a plan to build a suspension bridge somewhere between Shadowcreek and the Barnesville road. It would save hours of riding to check on livestock and go to fires. The ranger found a soft spot near a rivulet, laid down and was soon fast asleep.

In the morning, Buck remembered his personal promise to visit Mrs. Rust to see how she was. He still felt somewhat responsible for her sad situation. He hoped she hadn't lost everything. Back on the road to Barnesville he rode east and remembered where to turn for the house.

Mrs. Rust answered the door and for a moment Buck didn't recognize her. She had fixed herself up with a new hair style, new dress and shoes. She looked twenty years younger than a year ago.

"Well, hello Mr. Ranger. Come on in."

"Ah, ma'am, you are Mrs. Rust, right?"

"I used to be. Now I'm Mrs. Jack Gavin. My husband isn't here right now. He's away on business but my son is in the back. Shall I get him?"

"That won't be necessary, Mrs. Gavin. I happened to be in the neighborhood and wanted to see how you were doing. Guess I didn't need to worry."

"We had a real rough time at first. Mr. Gavin was a friend of mine in school, years ago. He recently lost his wife and happened to see in the paper where the government had filed a claim against Mr. Rust's estate. He came calling and admitted he had always been attracted to me. The

Chapter Three - 1907

result is what you see. He paid off the government claim so I could keep the house and property. We plan to fix it up really nice next year."

"That's great, Mrs. Gavin. I'm really happy for you. Hope you have a wonderful future."

With that Buck turned to go but Mrs. Gavin insisted they drink some tea together and discuss other topics. Buck complied and thought to himself that one never knows the potential of strangers. He bid Mrs. Gavin adieu and rode off with a smile. Out came his harmonica for the first time in weeks.

Stopping off at the Pitman house, he learned that Leo was still rounding up sheep. Buck told Mrs. Pitman to have her husband call the supervisor's office for an appointment. The supervisor was Kent Bolton and he would have all the maps, information, fees and permits needed. With this he rode south to report to Jug Handley and to see his boss. He needed to get back to the field and start work on homestead claims, timber volumes and a dozen other things.

Buck filled Kent in on what had happened. Kent sat there and chalked up another successful adventure to his ranger's power of persuasion. He agreed to let Buck know the outcome. The supervisor handed the ranger two new books—a new edition of the *Use Book* dated July 1, 1907 and a 42 page book titled *The Use of National Forests,* dated 1907. The latter one was for the public. It described the why, how, what, when and where of the national forest system. The *Use Book* had expanded to 248 pages including the appendix and index. Kent handed Buck several copies of the smaller national forest use book to hand out to the public. It had a red cover and was similar in dimensions to the thicker green covered *Use Book*.

Buck phoned Lakefield station and Luke answered. Alva and he had been mapping more potential agriculture areas and working on boundary surveying and had installed two more fire caches. The last one would be placed at Morgan's Meadow. Alva was up at Green Lake, trying to catch supper.

Buck stayed in Shadowcreek that evening and had his meal with Edna. He told her about Mrs. Gavin but didn't give a blow-by-blow detail on his run-in with Mr. Pitman. There was no need to worry her unnecessarily.

For the next two months Buck, Luke and Alva worked on homestead claims. Buck knew that most of the land had been surveyed and platted by contractors for the General Land Office in 1876. It didn't take the three men long before they realized the contractors had been paid for a fraudulent job. In some townships they couldn't find a single section corner and the topography on the plats was extremely inaccurate. It was doubtful that any field surveys had been made, even though well-drafted plats were filed. The various types of survey markings were wooden posts, rock mounds and blazed trees. If the tree happened to be exactly at the right place it was a corner tree, otherwise it was blazed as a bearing tree. Even with poor GLO markings, the ranger couldn't correct or change them to the proper locations. This was why so many sections were not a perfect one mile by one mile. For the most part, the recommendations on the homestead claims were cut and dried; they either could support agricultural growth or they couldn't. For the few that were close calls, they leaned in favor of the claimant, knowing that some wouldn't make it.

Alva understood about the funding of his position as a guard. He did go to several fires and earned money to help Buck's and his expenses. He figured it would be next spring before the ranger exam was offered.

In October, Buck called the office from Lakefield and asked Kent if Leo Pitman had complied with his sheep grazing agreement. Kent said Leo came in and was given a billing for the original trespass. Leo pled lack of funds so they worked out a time payment plan. However, he did pay the complete amount for all animals he was going to graze on his assigned district. The number of animals had matched that which Buck had been told. Kent mentioned he didn't think there would be any more trouble.

Buck told his boss they had completed the homestead claims and would proceed to other work. He wanted to check on some potential administrative sites and the possibility of building a bridge over Shadow Creek. Kent hesitated to ask questions about the bridge because he didn't know the answer—yet.

The three intrepid woodsmen rode directly to Morgan's Meadow. They surveyed the potential guard station site and made drawings as to a possible layout of the improvements. They also installed and stocked the sixth and last fire cache box.

Next they scouted for a narrow solid crossing to place a suspension bridge over Shadow Creek. After an hour of searching they settled on one that fit all their requirements. The water was only 20 feet wide and, although deep, there was at least three feet between the water level and the ground. It had solid rock on both sides for the cable anchor plus easily made level spots for the towers. The trail from Morgan's Meadow was east of the site and another trail from Shadow Creek to the railroad

Chapter Three - 1907

track two miles to the west and then to the Shadowcreek road close to the Dalton Ranch.

Buck sketched out a bridge plan. He knew there wasn't money to build it that fiscal year but he was always planning ahead for potential projects.

At this point Buck told his companions he wanted to check all the range districts on the east side of the Lakefield District which meant the land between the creek and the road to town. He wanted Luke and Alva to go south to town and head north, checking the number of animals, the correct brand and making sure the stockmen didn't trespass in areas where their cattle weren't supposed to be. They were not to confront anyone if they encountered an illegal situation. He would go north and start south. They would meet in five days at Harris Dalton's ranch. Harris was friendly to what the Forest Service was trying to do on the grazing lands and he had a few cattle of his own on one district. Buck also had another reason for visiting Harris.

There was a huge amount of land to cover in five days. Short nights and long days were the norm.

On the afternoon of the fifth day, Alva and Luke arrived at the Dalton ranch. They explained what they had been doing and that Buck should be along anytime. Harris said they could rest their animals in the barn and corral and stay the night if they wished. It was dark when Buck arrived. He had been slowed by several ranchers wanting to talk about the Forest Service. It was too late to eat in the house. Mrs. Dalton had finished cleaning up supper by the time Buck arrived. They ate outside and slept under the stars.

Harris had placed a For Sale sign at the entrance to his ranch so it could be viewed from the road. Buck had seen the sign when he was riding south to town after his run-in with Leo Pitman. The ranger had thought about the situation all that time.

Harris Dalton was a jolly man with a full crop of hair and swarthy complexion who worked all his life and made it pay. He could have passed for a person 20 years his junior.

Buck asked Harris, "Mr. Dalton I noticed you have a For Sale sign at your entrance."

"Yep! The Mrs. and I are getting up there, our kids are gone and we decided to take it easy."

"May I ask exactly what is for sale and how many acres you have, plus, of course, your asking price?"

"Well Buck, I have a house, a barn, two privies, a corral and three springs. The 28 acres are fenced, under cultivation and seeded with Timothy grass. I'm asking $1,700."

"I don't want to pry into your business but has anyone shown an interest?"

"Yep! But when they hear the price, even though they say it is reasonable, they don't have the money in one lump sum and that's how we want it."

"Well, Harris, what if the Forest Service showed an interest in buying your place for a guard station or even a ranger station. Would you be willing to sell it?"

"I'm willing to sell it to anyone who comes up with all the money."

"Now don't get your hopes up. I know we don't have the funds to buy it right now and even if we did, the supervisor and others would have to approve it. It might take some time or maybe never. I personally think it's a good deal for both of us and I'm going to lay it on the line with my boss."

"That's fine, Buck. It's been on the market for a year already and we're in no hurry." The two men shook hands and Buck thanked him for his hospitality. They headed south to Shadowcreek looking forward to some good food. Edna wasn't home so Buck left a message saying he would see her on his return from Harmony. He left for the office in late afternoon by himself. The other two rode for Crescent Ranch.

The next morning Buck and Kent got right to business. First the ranger reported on the range districts to the east side. They hadn't been able to count each animal but thought there were no major problems in that area. Some cattle weren't where they belonged. They didn't read signs and wandered about unrestricted. Salt blocks were not maintained very well and there were no line or ranger riders to keep the animals moving.

Next Buck showed the map and layout of a potential guard station site at Morgan's Meadow. If there wasn't enough money to build one, he believed it would be a perfect place for an improved campground. Then he presented Kent with the plans to build a bridge across Shadow Creek, since it was all Forest Service land and was sorely needed. He had worked up the plans and cost estimates. Kent asked a lot of questions and was satisfied with the answers. He thought Buck might be biting off more than he could chew, but realized by having a bridge about halfway between roads would save countless hours of administrative time. Finally Buck presented the possibility of buying a full built and maintained guard station or ranger station at Harris Dalton's ranch. He listed the reasons why it was too good a deal to pass up.

Chapter Three - 1907

At this Kent told his ranger he appreciated the effort he had put forth on the last two items but it seemed that every time Buck came in his office it cost money. He didn't like to keep saying no because he agreed with all the arguments, figures, maps and reasons presented. All he could do was pass them on to Washington with a cover letter supporting the projects and see what happened. That's all Buck wanted to hear. He asked if Kent had anything for him that was urgent.

"Well Buck, since you're caught up on the homestead claims, keep working on the timber volumes and agriculture lands but concentrate on mining for awhile. We have a ton of them to check on the forest."

"All right Kent, that's what we'll do. By the way, the six fire caches are in place on the district."

"Great work. I've already assigned the other three districts to do the same. We've been real lucky on the Maahcooatche for the past three years on fires. Due to deep snow and cool summers, we've kept the burned acreage way down."

"Sure have. I've been thinking about building a lookout platform on top of some trees, for a better view of smokes."

"Please Buck, don't ask me for more money. It's a darn good idea and it probably would pay for itself on one fire found early but the Washington office doesn't see it that way."

"I was just throwing out my idea. If we go ahead with it, there won't be a penny of government appropriations used unless you want to call our time money. Is that it?"

"Yeh! Just keep issuing small timber sales, check mining claims and do the inventories. Also, tell your brother that the next ranger exam is in April, 1908, here at Harmony. He's on the list."

"Great news and thanks for hearing me out, Kent."

They shook hands and Buck left the room. As he passed by Lucy, he gave her a little wink and a big smile, humming a tune as he left the building.

The meeting with Edna was short. Buck had to get to Crescent Ranch where the three had agreed to meet that evening. He stopped to say hello to Peter Blodgett and asked if there was any problem with the telephone poles or lines. There weren't.

It had been two years since he had talked to Harvey and Felicity Southcott, Luke and Sarah's neighbors. They were both extremely glad to see him and Felicity still poured thanks on the embarrassed ranger.

Sarah had supper ready so the three men, two boys and lady of the house had a rousing meal with plenty of stories and laughs. It was like old times in the Parley household.

Buck reported on his meeting with Kent. The telephone rang. Alva was requested to report to Tim Westgate for a fire on District 1. He took off immediately, riding east in the dark. The other two men turned in. Tomorrow and the rest of the year were to be busy.

1908

An Office Detail

I got a little detail
To the Supervisor's shack,
And I hadn't lit in Harmony
Till I wished that I was back
On the far end of my district,
Counting stock or building trail,
For to work inside an office
Is like doing time in jail.

This bending o'er a table,
And a writing all the day,
Is a-making me hump-shouldered,
And my hair is turning gray,
It shore will be my finish
If they don't relieve me soon,
For my bewhiskered, sunburnt features
Is getting paler than the moon.

I thought that I had troubles
When on my district all alone,
But I've found that serious trouble
Was a thing I'd never known.
When I git back on my district,
You can bet your life I'll stay,
And be thankful to my Maker
I can draw a ranger's pay.

James H. Sizer

CHAPTER FOUR

1908

Buck and Edna were dancing up a storm. Both were passionate music lovers so when Founders Day was celebrated in Harmony on the second weekend of March, they made arrangements to stay Saturday night: Buck at the boarding house and Edna at the home of her friend, Alice O'Neil. It seemed as if half the town of Shadowcreek and the village of Fish Cut participated with the good folks of Harmony. The afternoon was filled with the local brass band playing marches, waltzes and patriotic tunes in the gazebo of the town park. An outdoor banquet was prepared for supper. The mayor and councilmen made speeches and the wine, beer and hard liquor flowed freely. It was almost midnight and neither Buck nor Edna showed any sign of fatigue. This was their first Harmony's Founder Day party. They promised it wouldn't be the last.

Someone called out Buck's name. He turned and it was his boss, Kent.

"Got to talk to you Monday morning," he said. "About some important news and you're involved."

"OK, Kent I'll be there. Having a good time?"

"Yea! I'm trying to escape from my problems for a few hours." Kent noticed the ranger was with Edna. He went on. "I'm amazed you got this big oaf to take things easy on a weekend. He usually doesn't know what day of the week it is. Every day is work day."

"Well, Mr. Bolton, I've almost given up trying to make him slow down. What do you think I should do?"

Edna winked her eye as she said this and Kent caught on. Buck was squirming as the talk centered on him. He didn't catch the wink.

"If I were you," Kent went on, "I'd either give him a small daily dose of arsenic so he would have to slow down or if you didn't want to take so long waiting for it to take, then I'd speed it up by hitting him over the head with a two by four. We have a good hospital in Harmony and he could take his vacation as a pampered patient."

"I think both those ideas are commendable," replied Edna. "I could pamper him at my place instead of the hospital."

For a moment Buck was caught off guard. Then realizing they were teasing at his expense he countered, "Your ideas won't work Kent. I've practiced hitting my head on a two by four every day for twenty years and have acquired this extremely hard noggin."

"Hard head is right," said the supervisor. Edna agreed. The three of them parried in this manner for awhile. Finally Kent called it a night. Buck said he'd be in the office on Monday morning, first thing.

Kent yelled back to Edna as he left, "I'm counting on you to take care of that big lug. This forest would be in deep trouble without him." Edna replied that she certainly intended to, even if Buck didn't like it.

Kent and Buck huddled in a closed door conference most of the morning. The supervisor ordinarily called all four rangers in about this time of year but he had an emergency situation and sent a list of additions and changes in work by mail earlier. He had another reason for Buck to be there.

"Buck, as you know, I've depended on you for many things on this forest that I haven't asked the others," said Kent.

"I guess, if you say so, Kent," replied Buck.

"Well, I'm going to ask for help again. I received a phone call the other day from my brother asking me to come home right away. He said our folks are not expected to live much longer. They have a variety of health problems and now they have pneumonia. They live in Flagstaff, Arizona, so I figure to be gone at least two months. It's a situation somewhat like that first year of yours."

"I'm sure sorry to hear about your parents. I can sympathize with what you're going through."

"Thanks. Now here's the problem Buck. With Garth gone and me out of the picture for a couple of months, this forest needs a supervisor. I laid out my problem with the Washington office and they are sending a man named Sanford Picton who will be acting Forest Supervisor in my absence. Now what I'm going to say must stay within the walls of this room. I've never met Sanford but I do know some friends in high places in Washington. I asked them about him in an unofficial capacity and if he would do the job. Every answer I received was yes he could do the work if he was sober. He evidently works somewhere in the forester's office. This will be a detail for him to see how the field offices operate. I then asked why he hadn't been fired or let go. The answer was that he was a nephew of a high ranking senator and as long as he didn't commit

Chapter Four - 1908

a crime, he would remain. He's expected to be here Wednesday on the train. I've got to leave tomorrow for Arizona. So you see Buck why I need you?"

"Well, Kent, not exactly. Of course you can depend on me to keep all this to myself but I don't understand just what you want me to do."

"In a nutshell, I want you to be his shadow. I'll relieve you of your field duties 'til I return. I'm sure Luke and your brother can do what is necessary in your absence. I'm not looking to get Sanford fired or anything like that. I'm just worried about the forest operating on a rational and dependable basis. I think the best way to do it is to make you acting deputy forest supervisor so that you can be his shadow without raising suspicions. I've composed a letter in long hand and signed it."

Kent gave the letter to Buck. It read:

Dear Mr. Sanford Picton,

Welcome to the Maahcooatche National Forest. It is my understanding that in my absence you will be acting forest supervisor. You will be working closely with Buck Stonewall, ranger of the Lakefield District. During this time his title will be acting deputy forest supervisor. We recently lost our full-time deputy forest supervisor to another forest. To administer the forest properly, we need both positions filled. Buck will work mostly in the office with you. The everyday work on his district will be done by his guard Luke and his brother Alva. The latter will be taking the ranger exam next month. They are both extremely competent men.

Do not hesitate to ask Buck to do anything you want or need. He is here to serve you and the forest. He will make speeches or greet dignitaries or inspectors. He will also act as your guide as you travel around the forest. The other rangers are aware of the situation. See you when I return and good luck.

Best wishes,
/s/ Kent Bolton
Forest Supervisor

The ranger laid the letter down.

"There it is, Buck. What do you think?"

"My God, boss, you want me to be a baby sitter to a grown man," exclaimed the ranger. His reaction told the supervisor his ranger was not happy.

"Buck, I'm pleading with your sense of loyalty to the outfit. If you won't or can't bring yourself to do this for a couple of months then this forest could be damaged for a long time to come. I'm just not sure what Sanford might do if left alone to make all the decisions. I'm certainly not going to order you to do it. I have too much respect for you. If you want to think about it for a few hours and tell me this afternoon, that's fine. Whatever you decide."

"I'm sorry I raised my voice, Kent. It was a knee jerk reaction. Of course I'll do it. Please don't worry about anything while you're gone. You'll have too much to think about at your folks' home. In case of a real emergency, would you leave your telephone number and address with Lucy?"

"Sure will. You're a life saver, my friend. I'll never forget you for this. As if I could anyway. I realize I'm asking you to do something above and beyond the call of duty. If you have any question about something you don't know about, be sure to call or write anyway. It doesn't have to be an emergency. If you have any extra expenses that the government won't pay, then I'll reimburse you when I return. Try to keep Sanford out of jail. If it happens, at least you tried. One more thing. Be sure not to enter any booze problem or what you did about it in your daily diary. You are free to write a daily report on your activities for my eyes only. Remember, I'll back you all the way on any decision you make. Let's talk about some new requirements that just came out."

Kent showed Buck the monthly fire report form 944. It listed seven requirements.
1. Cause and location of fire.
2. Area burned over.
3. Amount, kind and value of timber burned.
4. Amount, kind and value of timber damaged.
5. Damage to reproduction.
6. Cost to Forest Service, including cost of forest officer's salaries, tools, etc.
7. Summary of evidence regarding the origin of fire with names and addresses of any witnesses.

Next, Kent showed Buck a memorandum asking for information on all water development possibilities on the district. The forester would consolidate all reports for the President, to be used in formulating a national conservation policy. The report would cover each stream that could produce 100 horsepower or more.

Kent asked his ranger to prioritize and have Luke and Alva do the job soon. Buck replied that Luke wasn't even on the payroll yet for the fire season. At this, the supervisor phoned Crescent Ranch and asked for Luke. His son Matthew said his father was gone for a few minutes but would have him return the call.

The supervisor mentioned the office received books from the Forest Service library. All rangers were to study them diligently, said the memo. Buck said it sounded as if the Washington office wanted the on-the-ground personnel to also start on-the-job training. He said it would be something he would do on a freezing winter night with only a candle for light. The supervisor knew his ranger was joshing, so he didn't respond.

The last important item that was discussed described an inventory of all property, due in the Washington office where all the property records were kept and managed. The memo said an inspector would come to the forest to brand and condemn all broken and worn-out tools so the forest could be relieved of accountability.

The call came in and the supervisor asked Luke if he would be willing to come to work right away instead of his usual May 1st. Luke said he could in three days, after he settled some business matters. Buck got on the phone and told his guard that he would call Crescent Ranch Thursday morning and go over the priorities for work.

It was almost noon. To show his appreciation, Kent bought Buck lunch at the Golden Restaurant and Saloon. After lunch the ranger told Zeke the bartender that he might be seeing him much more in the days and nights ahead. He didn't go into details.

That afternoon Buck sat at Garth's old desk where he would spend the better part of the next two months and mapped out how he would manage the Lakefield District from long distance. He would have Luke and Alva fix the downed telephone line from the station. It was now critical to make contact. Next he would give the two a list of streams he thought might produce 100 horsepower year around. They needed to double check his memory and add or delete when necessary. The property inventory would be easy. They only had Lakefield station and Crescent Ranch. They were to check for any more Homestead Act claims and, if possible, complete the district mapping of potential agricultural lands. The timber volumes were still to be worked on but it was too large a job to complete in the next two months.

Buck contacted Alva and asked him to come to the supervisor's office first thing Wednesday morning. His brother had been biding his time at the Harmony boarding house.

Buck telephoned the other three districts. Miraculously Tim, Ralph and Hugh were all close by and available to talk. They each knew that

Kent had an emergency and would leave the next day. Buck brought them up to date on his new title in the supervisor's absence and how he really needed their help. He said the acting supervisor Sanford Picton would arrive in Harmony on Wednesday. He would meet him at the depot and get him settled into a room either at the hotel or boarding house. He went over their current activities and what might be some problem areas in the next couple of months. They should expect Sanford to tour each district. Although he would accompany the supervisor, each ranger needed to make himself available when the time came. He urged them to keep the lines of communication open. The last thing Buck mentioned was that Kent had indicated there would be no ranger meeting in Harmony that spring. He thanked each of them for their cooperation and willingness. Buck was up to date on the districts. He had not told any of them about the acting supervisor's attachment to the bottle.

Kent was a bachelor. Buck asked if he could help him with his horse, close the house or even pack for the train the next day. The supervisor said if Buck was through for the day they could both ride together to the Harmony Livery Stable. John Roster was the owner of the stable. He greeted the two men and told Kent his horse Fiddler would be well taken care of. Buck asked about renting a horse for the new supervisor for about two months. John said he had just the horse. Buck told Kent he was going to pay for the horse's rental for one week. That way if he didn't get reimbursed by Sanford, it wouldn't hurt too much financially. He had figured a horse might be needed at the depot when the Wednesday morning train came in, supposedly with the acting supervisor on board.

The two men walked to Kent's house about four blocks from the stable and six blocks from the office. Buck had never been there. The supervisor's home was definitely not a typical bachelor's quarters. Even the outside had nice vegetation and had been freshly painted. There were two bedrooms, a kitchen, a dining room, a living room, three wardrobes and a hallway. Simple, but clean and comfortable. Kent had purchased the house in 1903 when he first came to the area. The two men talked about family and friends, as the supervisor offered Buck his usual whiskey on the rocks.

Buck's offer of helping to pack was turned down but the ranger insisted he would come by the next morning and carry one of the two suitcases to the depot. Before they separated, Kent handed the house key to Buck and said he was welcome to use it in his absence. At least it would save him some money. It had a telephone and was stocked with

Chapter Four - 1908

food. Buck was flabbergasted. He tried to change his boss's mind without success.

Events went like clockwork Tuesday morning. Kent and Buck walked to the depot; the train was on time; the two men shook hands vigorously and the acting deputy forest supervisor was now in charge for at least one day.

Edna was excited when Buck told her the news about his two month detail in Harmony. He said that even though they were only a few miles apart, the next days were critical to his staying in Harmony. If all went well, he would ride to Shadowcreek for the weekend or at least on Sunday. The answer should be known by Friday.

Back at the office, Buck and Lucy discussed how the office would operate. Lucy knew who to contact in Washington if things went awry. She gave Buck her home address and telephone number just in case. He in turn gave her Kent's home number which she already had. They were prepared for the worst but hoped for the best. The boss's calendar showed he was supposed to give a speech to the newly formed Rotary Club on Thursday noon. An inspector from the Washington office was to visit in April at the same time as the next ranger examination. Since he would be tied up with the exam, Buck asked Lucy to try and postpone the inspector's visit, explaining that the forest supervisor was called out of town on an emergency.

For the rest of the day and a couple of hours after Lucy left, Buck studied and wrote down names and potential problems that he might be faced with as reported by the other rangers. He didn't want to be blindsided with the unexpected.

Alva went to Kent's house early the next morning, after Buck told him where he would be living. They had breakfast together and talked about the farm and how Ivy, Henry and Charlie were doing. The latest letter was good news and there seemed to be a positive financial change.

Looking at the clock, Buck said, "We've got to be going to the livery stable. I'll ride Titus and you ride Sanford's horse Gin Rummy to the depot. The train arrives at 8:30."

It was 8:50 when the train arrived. Several people embarked but there wasn't a sign of Sanford Picton. They were about to leave when the conductor yelled out that there was a man asleep in his seat who was supposed to get off at Harmony. He asked if Buck or Alva knew him or were waiting for him. Buck said that they were waiting for a man they didn't know or even what he looked like. He asked if he was sleeping, why didn't the conductor just wake him up. The reply was that he was either drunk or dead.

"Uh oh," said Buck. "Alva you stay here and tie up the horses. I may need your help quickly."

Buck bounded into the passenger car and the conductor pointed. Sanford was out of it. An empty quart gin bottle was on the floor beneath the seat. He opened the window and called his brother to come get the luggage. He then picked up the limp body and threw him over his shoulders with a fireman's hold.

After the train left Buck decided to place Sanford stomach down over Gin Rummy's saddle and hold his legs while they walked to the Harmony Hotel. The luggage was tied to Titus's back.

Calvin Merton, the desk clerk, found a room on the second floor. He asked Buck to pay one night's lodging. Buck threw Sanford down on the bed and took off his shoes. They left the luggage unopened and withdrew. If the supervisor woke up, Calvin was to call the office and tell anyone who answered.

Buck, Alva and Lucy discussed what happened, concluding it was going to be a long two months. The ranger knew he would have to make the Rotary speech the next day.

Just at closing time, the phone rang. It was Calvin saying that Sanford was up and wanted to know where the closest bar was. At that the two men raced to the hotel and confronted the acting supervisor as he was about to leave. They introduced themselves and brought him up to date on what had happened. They asked if he would join them for supper at the Golden Restaurant and Saloon. At the mention of the last word Sanford's eyes lit up and he said, "yes".

Getting something into the supervisor's stomach was important. He finally came around, sober enough so as to apologize for his behavior. He guessed their first impression was not very positive. Buck kept the conversation going and brought the supervisor up to date, as well as could be done in a two hour time frame. They left, out the side door, away from the bar. Pointing out where the supervisor's office was, Buck and Alva took Sanford back to his room and said they would see him in the morning. At least that part was to come true.

Sometime past 1 a.m. Kent's phone was ringing off the wall. Buck finally got his thoughts together and answered. Zeke was on the other end and thought Kent should know what the new employee was up to.

"I'll be right there," said Buck. "Don't let him leave."

"In his condition he ain't going anywhere," answered Zeke.

Chapter Four - 1908

Five minutes later Buck entered the swinging doors and saw Sanford crumpled in the corner. Before placing the limp pile over his shoulders, the ranger asked Zeke if he would decline service for drinks to the acting supervisor. The bartender hesitated because it would cost him money. Buck said to keep track of each time Sanford came in and asked for a drink. Zeke was to refuse the request and later would be reimbursed for the phantom drink. This would mean a 100% profit. The bartender agreed but the ranger warned there would be only one drink each night. It didn't count if Sanford came back and asked every ten minutes.

Back in the hotel room, Buck waited for more than a half hour to make sure his boss was going to sleep through the night. He then went back to Kent's house and slept three more hours.

It was 10 a.m. by the time Buck got Sanford out of bed, dressed and finished with breakfast. They walked to the office. Lucy, Luke and Alva were there. Introductions were made and the supervisor was shown his office. Buck had made a summary report on what activities were taking place on each district. He also asked if the supervisor wanted to make the Rotary speech. Buck got the job. The boss closed his office door and made some phone calls back East. In the meantime, the three Lakefield District personnel discussed the list and priorities that Buck made earlier. Luke and Alva were ready to get going. They were late and wanted to get to the station that evening with the pack train full of supplies and equipment.

In about an hour, Sanford's door opened and he beckoned Buck in. The deputy could smell the whiskey on the supervisor's breath. He thought it amazing his boss could speak so easily without slurring his words.

"Buck, I first want you to know that I usually don't do what happened last night. Guess I must have left for the bar after you went to bed. I can only guess that you were called to take me home. It won't happen again. I'll stay alert enough to find my own way back to my room. Do you have anything to say?"

"Well Sanford, now that you brought it up, I'm personally out one week rent on your horse, Gin Rummy, as well as your first night at the Harmony Hotel. If you can't afford it right now, that's all right."

"No, no. I've got the money with me." With that he brought out a wad of bills and squared it with the ranger.

Harmony had more than one saloon. The closest one to the office was the Log Cabin Bar. They did not serve food which is why Buck and Kent rarely went there. At noon that day Sanford left and headed straight for the Log Cabin.

Chapter Four - 1908

In the meantime, Buck went to a private room at the Golden Restaurant and Saloon to give a talk to the Rotary and eat a free lunch. His talk was mainly on the many changes to the national forests coming from Washington with specific comments about the Maahcooatche. When he asked for any questions at the end of his speech, almost 100% in attendance shot their arm into the air. Some of the questions that afternoon were not pertinent but Buck didn't bat an eye. He explained what he knew in detail, including the "why." If he didn't know the answer he wrote the name of the person who asked and promised to get the answer to him. He threw humor into his answers and said he understood some of their frustration on government rules and regulations. Then the question came that Buck prayed would not.

"Mr. Stonewall, I understand Forest Supervisor Kent Bolton had to leave for a few weeks. I happened to be in the next room last night when you came in and picked up a drunk off the floor. Is he a friend of yours?"

Buck thought about his answer for a few seconds. Then he said, "I'm not going to lie to you folks. His name is Sanford Picton. He's from the Washington, DC, office of the Forest Service and will be here as acting forest supervisor until Kent returns. I will also be in the office to assist Mr. Picton. He arrived yesterday after a long train trip. My brother and I ate supper with him here and then left him at the Harinony Hotel. Being in a strange town, Mr. Picton wanted to meet people and make some friends. Not knowing his way around, he returned to the saloon here and proceeded to have a few drinks. Zeke the bartender called me to ask if I would help take him home. That was it. There's nothing more to tell. I'm sure most of us at one time or another have had too much to drink, especially when we were lonely and in a strange place. Any more questions about the Forest Service?"

There were three more questions and the meeting terminated but not before Buck was given a long, loud applause. On the way to the office, he hoped his answers had corrected some wrong ideas and been a positive influence to the Rotarians.

The supervisor and deputy supervisor worked separately that afternoon. Just before closing time, an upset Mr. Sherman Clapham came barging in demanding to see the supervisor. Buck overheard Lucy asking the man to wait and she would see if he was available. She looked in the supervisor's room and promptly shut the door. Motioning for Buck to come into the reception area she introduced the man to the deputy. She said the supervisor was unavailable at the moment. Mr. Clapham was then escorted to Buck's room where he was cordially asked to take a seat and say what was on his mind.

Still angry, he sputtered, "I was turned down last year by your office in Canyon Springs and I want to know why."

Buck knew that he was talking about Tim Westgate's District 1, with headquarters at Canyon Springs. "You were turned down for what, sir?" asked the deputy.

"Turned my entire homestead claim down; that's what they did. It was perfectly legal and it isn't fair. I'm mad as hell and I want some answers."

"Did you bring anything to show why they turned you down?"

"Nope! Don't you believe me?"

"Of course I believe you were turned down, but I need some more information so I can find out what happened. Can you give me the legal description of the claim?"

"Nope! You really don't believe me do you?"

"I believe you but I need to know where it is, Mr. Clapham. Can you show me on this map?"

"Right there on the map. Now what?"

"How can I reach you by mail or telephone? I need to talk with the ranger on the district and find out the problems. I'll do it right away."

"You have 'till 9 a.m. tomorrow. I'll be here and I don't want any runaround."

"See you tomorrow." With that Buck got up, escorted Mr. Clapham to the front door and bid him adieu.

The deputy tried to get Tim on the telephone but was told he wouldn't be in until tomorrow morning at 8 a.m. He then asked Lucy if there was a folder on Mr. Clapham's claim. She found one but it contained only the claim, not the reasons for denial. He looked closer at the original claim and saw several things that bothered him. It was for 360 acres. It was dated two months earlier at the top than the date at the bottom signature. Looking at the map, he thought the legal description was a spot set aside for an administrative site. He also noticed it contained a lake of more than 50 acres. Too many unanswered questions, he thought.

Lucy left and locked the front door. Buck checked on Sanford who was sleeping at his desk. Out loud he said to no one in particular, "Geez, I don't know if I'm up to doing this for two months. It's only been a day and a half. Maybe if we can go on a forest show-me trip next week he'll sober up. Got to do something."

Chapter Four - 1908

The phone interrupted his thoughts. It was Tim. He had come back to his office unexpectedly and heard Buck was trying to get him.

"Thanks for returning my call, Tim. I just had an irate customer of yours by the name of Sherman Clapham come in and complain about the denial of his homestead claim. What can you tell me?"

"Yea! Mr. Clapham is a hot head. Well, Buck, we surveyed his claim and denied it for several reasons. First of all, it was well over the 160 acres limit. Secondly, it was on land that was already approved for homestead. You'll note he showed an earlier date trying to cover that up but it was bogus. Thirdly, part of the acreage is set aside for an administrative site. Fourthly, he admitted the claim was just to obtain ownership of the lake so he could turn it into a commercial venture. And lastly, at least a quarter of the acreage is nothing but rocks. There's no way it can grow anything."

"Just as I thought, Tim. Did you spell this out in writing?"

"Sure did and he has a copy I gave to him personally."

"Thanks for the information. You may see me with the acting supervisor next week. Will you be available?"

"Yup! You can catch me up when I see you. Hang in there."

"I will unless I hang someone else first," Buck quipped and they both laughed.

That evening the two forest officers ate supper and separated, bidding each other goodnight. Buck headed for the Log Cabin bar and made the same deal with Ned; the bartender there, as he had with Zeke. It was going to cost him money but that was better than packing his boss home to bed every night. If he wanted to drink himself to sleep every night in his hotel room that was also fine. He went to his temporary home and called Edna. Hearing her voice relaxed him and he talked for a long time but not about his work problems. Edna sensed Buck wanted to let off steam, so she just listened. When he said he wasn't sure about the weekend, or any weekend for the next two months, she said, "I'm coming to Harmony Saturday morning and staying at Alice's house Saturday night. We can talk and maybe go to church on Sunday."

Buck was feeling much better as they said several good nights.

The next morning the clerk and deputy were on time. The supervisor had not shown up. Promptly at 9 a.m. Sherman Clapham walked in and went directly to Buck's room without saying a word.

"Good morning, Mr. Clapham. How are you?"

"Well, what about my claim?"

"I've looked at your claim and talked to the ranger, Tim Westgate, and he tells me you received a letter from him explaining why you were turned down."

"What if I did. It was full of lies and I tore it up. I want some answers."

"OK then, let's go over each item of denial. First, the act allows no more than 160 acres on a claim and you wanted more than twice that."

"So what. I'm willing to cut it to 160 acres if I can pick the area."

Buck wasn't going to argue. "Secondly, you have a date discrepancy on your claim. A date two months earlier than when you signed it. Between those two dates another person put in a claim for some of the same area and it was approved. It looks as if you predated the claim when you signed it."

"You can't prove anything. I just made a mistake. The signature date was wrong."

Buck went on. "Thirdly, part of the acreage you claimed had already been set aside for an administrative site. An administrative site may be recreational sites, a guard station or a number of other possibilities. But, let's go on. Fourthly, the ranger said you admitted planning to use the lake for commercial purposes. That's not the way to do it. Your claim didn't show that. And fifthly, a large portion of the acreage is not fit for agricultural growing unless you plant a huge rock garden."

"I can always cut that area out. If that's all the reasons you have, then I still don't think it's fair. I made some minor errors and you guys won't let me correct them."

"I'm afraid it's too late now, Mr. Clapham."

"I want to take this to Washington, DC. Who does my attorney send the letter to and what's his address?"

"That's your right. I'll have Miss Neville write down what you asked for. Is there anything else I can help you with today?"

"Nope! The sooner I get out of here, the better I'll like it. Sure wish we were back in the good old days with no government interference." With that Sherman left the building with the name and address he requested.

There was still no sign of the supervisor. Buck wasn't going to check. He could be anywhere. The telephone rang. It was Ned at the Log Cabin wanting to know if Buck's boss asked for a drink in the morning would that count for the night request also. Buck didn't answer but said, "Don't give him a drink under any circumstances." He ran out of the office to the bar. Sanford was sober but he was raising hell for not being served. Buck finally persuaded him to come to the office.

Chapter Four - 1908

Back in the office, Buck asked if he could set up a forestwide show-me trip for the supervisor, starting Monday morning. They would travel by horse and one pack mule. It would be a fast moving and intense trip of about two weeks. Sanford liked the idea and told Buck to make the plans. That afternoon the deputy called Tim at Canyon Springs and said they would be there Monday afternoon. He should be ready to show the district improvements, projects and problems. Buck also wanted Tim to ride with them on Thursday morning to Ralph Hempstead's district office at Fulton, explaining that he had never been there himself. He asked the ranger to obtain food for three people while on the district.

Buck called Ralph, explained the situation and asked the same as with Tim. They would be there on Thursday and travel to Hugh Tanner's headquarters at Barnesville on Sunday. The same information was given to Hugh. Buck would show Sanford his own district starting the following Wednesday and they would be back in the office by Saturday.

It was a calculated timing risk, but worth the effort to get the supervisor out of the office for two weeks. He remembered Kent said he would owe him big time if he kept Sanford out of trouble. Buck decided to keep a running list to present to his boss after his return. He smiled when he thought about the fun he could have.

The rest of the work day went smoothly. All visitors were handled competently by Lucy. She seemed to know the forest well. Buck vowed he would never be tied down in an office again. He would much rather work 12 hours a day, six or seven days a week in the field, than for eight hours in an office. Many years later he would change his mind but wouldn't limit his time to only eight hours even then.

Buck telephoned Edna and asked if she could be there the next morning. He needed a break from the supervisor but didn't dare leave Harmony, in case there was trouble during the weekend. It was rotten that he had to have Edna do the traveling. Buck never felt as trapped in a job as at that moment. At other times, he would just move on. Now he felt an obligation to Kent and a responsibility to the Forest Service. His sleep that night was fitful. At least no bar called at midnight to come get his boss. He had visited the three saloons in town with the same promise to cover the cost of a phantom drink for Sanford.

Edna showed up at Kent's house well before noon. She knew immediately that something was troubling Buck. After some small talk, he finally asked her if she would listen to his story that seemed to have no solution. He didn't want to burden her with his crisis but didn't know who to turn to. Would she hear him out? Of course she would. Starting with his promise to Kent, meeting the train, the drinking binges, the phone calls, keeping the office running, being questioned by the people at the Rotary club and now leaving for a ten day to two week show-me trip, he couldn't see how it could go on for another six to seven weeks. Edna had never seen Buck so down or so vulnerable. If he was this way after only a few days, she shuddered to think what would be the result of two months of hell. Her mind was racing and her feelings for Buck demanded that she find a solution immediately.

Suddenly Edna had an idea. She asked, "Is Sanford married or single?"

"Single, why?"

"Has he ever mentioned women or a woman?"

"One time he told me his drinking problems started when the woman he loved left him for another man, just before they were to be married."

"Buck, I've got a plan. There is no guarantee it will work, and if it doesn't, we'll try something else. I want you to go find Sanford and stay with him until this evening. Don't let him take even one drink. I'm going to see my best friend Alice O'Neil and explain that we need her to make a foursome for dinner at the Golden Restaurant. We'll meet you there at 6 p.m. You introduce Sanford and I'll introduce Alice. Explain beforehand to your boss that the woman he is having dinner with is a widow. She is not against anyone having a drink now and then but she abhors drunks, loud mouths and show-offs. She rarely goes out socially and is very independent."

Buck held his tongue at that remark but couldn't help thinking it reminded him of Edna.

"I have my work cut out for me in persuading Alice to go tonight. I know she likes you and if I explain without going into details that Sanford is your boss and it would be a great favor to you if she would be his dinner partner, she might do it. I know she, wouldn't do it if we considered it a date. We have to work together on this Buck. It won't be easy for either of us. What do you think?"

"I'm desperate, Eddy. Let's try it. If it doesn't work, then I can always use chloroform when he isn't looking." Buck gave a little snicker when he said it.

They ate lunch at the house, then hugged and kissed and parted with a good luck comment.

The deputy supervisor knew he had to work fast. Already he might be too late to stop his boss from taking several drinks from the bottle in his room that morning. There was nothing he could do to prevent Sanford

Chapter Four - 1908

from acquiring a bottle from the store. Buck knocked on his hotel room door and there was a grunt from inside. He identified himself and in a moment the door opened and there stood the supervisor, not drunk but well on his way.

"Come on in, Buck," said Sanford. "This town doesn't like me. I couldn't get a drink last night at any of the bars. They all said I'd been banned. Can you get them to change their minds?"

"I've come here, Sanford, as a friend, not a deputy. I'm going to talk to you as a brother, so I want you to listen very hard and give me an honest answer. Please sit down in that chair and don't ask anything until I'm through."

"You sound really serious, Buck. Is something wrong?"

"Not only is nothing wrong but you're in for a big surprise this evening. You may thank me before it's over. Now, here's what it's all about."

The ranger went on about the four of them having dinner. About Alice's aversion to excessive drinking, braggarts and big mouths. He explained how she was a widow who was very selective with whom she keeps company. To make it work, however, Sanford must not take another drink for the day or evening. He must sober up, clean up, shave and wear his best outfit. They would stay together until it was time to walk to the Golden.

"Well, Sanford, what do you say? I think we'll have a great time."

"What does she look like, Buck?"

"She's very appealing. In fact, I'd say she is extremely good looking. She's about three inches shorter than you, has dark hair but don't ask me the color of her eyes. I just don't know. The only time I met her was when Edna and she were together and I was looking at Eddy practically the whole time. She seems very intelligent with a sense of humor."

"Gosh! I'm not going to turn this down, Buck. I can hardly wait. I've forgotten what it's like to have a good conversation with a lady."

Sanford washed, shaved and changed into his best clothes. Buck asked him to come to Kent's, where he would do the same. By 5 p.m. the two gentlemen were ready for a pleasant night out.

Edna was successful also. The men arrived early and at 6 p.m. sharp the two ladies were escorted to Buck's and Sanford's table. Buck thought Edna was beautiful in her dark blue linen skirt, dotted Swiss organdy leg-of-mutton[14] sleeved blouse with a stand collar and white lace jabot.[15] She wore a white, wide brimmed leghorn[16] hat with pheasant feathers. Alice's light blue calico dress had a pleated yoke and a wide sash with contrasting colors around her waist with a bow at the back. Her hat was a straw boater with silk flowers. Both ladies wore pointed toe shoes with a French heel. Edna's was a high laced white shoe and Alice's a high buttoned black pair. Introductions were made and four glasses of sherry were ordered. The supervisor sat across from Alice. He couldn't keep his eyes away from her. He was quiet, reserved and completely smitten. As the evening progressed, Buck and Edna smiled and winked at each other several times. Alice gave no sign she thought Sanford was anything special, although she offered her share of the conversation and laughed at his stories about life in Washington.

Near the end of the evening, Buck suggested something he never thought he would. "Since tomorrow is Sunday, how would everyone like to meet at church for a sermon and some singing?" The other three readily agreed. The women went their way and the men went to the Harmony Hotel.

Once inside the room, Sanford exploded. "Buck, this is the best day of my life. I'm in love with Alice. I'm not kidding. I couldn't take my eyes off her. She is beautiful, refined, gorgeous, intelligent, captivating, humorous, elegant and graceful."

"That's great but most of what you said is redundant. Now Sanford, settle down. You've just met her and by your own admission it's the first time for awhile you've been out socially with a woman. Whatever you do, keep your feelings to yourself. At least until she gives a little something that indicates it might be mutual."

"I will, I will. I don't want to ruin it, Buck. What should I do?"

"All I can say is to keep yourself reserved and low-key. Let nature take its course even if it takes a long time. Don't hurry anything and above all don't take another drink. If Alice ever finds out you've gone on a bender or aren't stone cold sober, then you can say goodbye to her forever."

"I'm on the wagon starting right now. That one sherry tonight was my last drink for a long time. I'm not going to buy any liquor for our trip either, even if it takes a couple of weeks."

"That's great, but I think we might have a wee sip now and then as we sleep outside on these cold nights."

They parted but not before Sanford handed Buck three full bottles of booze plus a partially filled one.

Buck whistled and hummed all the way back to Kent's place. He placed the four bottles in the closet.

The next day, the group attended church and walked in the park. It

Chapter Four - 1908

was a gorgeous day. Happiness was in bloom. Buck and Edna let Sanford and Alice go off by themselves. Curiosity got the better of Buck, so he asked what Alice had thought about his boss. Edna explained that she enjoyed the evening and thought Sanford was very nice. She liked his manner and his reserved disposition. Otherwise there would have been no church meeting or walk in the park.

"Remember Buck, Alice is very cautious. She doesn't want to get hurt. Your boss was very wise to act the way he did. I hope he can keep sober and not barge in with his feelings. Yes, I saw how he looked at her and knew he was hit hard. What did Sanford say?"

"He told me he was in love with her. Eddy, wouldn't you say that was ridiculous loving someone at first sight, without getting to know them?"

Her face flushed. "Well, it is possible, I suppose, to fall for someone right away without knowing them. Sometimes I guess a person has a deep feeling about someone else, knowing that person is destined to be their soul mate. Haven't you ever thought you were in love with somebody after first meeting them, Buck?"

"I don't think so. I guess it might be possible."

Edna thought better of pursuing the subject. "I have to start back to Shadowcreek soon, Buck. Guess we won't see each other for a couple of weeks."

"Yea! Darn it. I thought we'd be together more during my time in Harmony, but it doesn't seem to be turning out that way. Anyway, I'm going to ride with you back to your home. We have time and I've got almost everything packed and ready for our show-me trip."

"You don't have to do that, Buck. I can take care of myself."

"I know you can, Eddy. You're the most self-reliant woman I've ever come across. But I really do want to ride with you. We seem to have so much to talk about and we both enjoy the country road between the towns."

They excused themselves from Sanford and Alice, after making sure it was all right. The trip to Shadowcreek was not without incident. As they approached Fish Cut, a commotion near the confluence of the two bodies of water caught their attention. People were shouting and running about waving their arms and hands. Buck thought they should investigate and as they approached the wild group, he realized a dog was in the Crescent River. There was some debris held fast that the dog had hit and was now mashed into. The current was so swift it couldn't escape either backward or to the side.

The ranger immediately got his lasso, tied one end to Titus, stood on the shore and heaved the rope as far as he could. It missed, but Buck realized he would have to throw it above the dog and let it float down. At that moment he would tighten the loop and try to grab the animal in any manner he could. After the fourth try, Buck still wouldn't give up. The dog couldn't keep it up much longer. Every throw resulted in the rope touching the animal in some way but not with a loop so it could be brought ashore.

All this time the village folk were still yelling but now they were encouraged by the ranger's presence and attempts to save the canine. On the fifth throw, the landing was five feet from the animal. At the moment the rope touched the dog, Buck gave his end a wave action whip. This carried all the way to the loop which made it lift enough to go over the animal's head. He immediately pulled it taut and whistled for Titus to move inland. The crowd cheered and the exhausted dog was rescued. They shook Buck's hand and patted Titus on the neck. The owner of the dog wanted to pay something but the ranger thanked him and declined.

The journey to Shadowcreek continued with the two riders thinking wonderful thoughts about each other. Buck hadn't wanted to say goodnight, so when he got back to Harmony it was late. After feeding and brushing Titus, he hit the bed and was asleep in five minutes.

* * *

It was a five hour ride to Canyon Springs Ranger Station. The hum of the generator could be heard as they approached. The station not only had electricity but a large barn, a separate office, two residences, two outhouses, a corral with three horses and three mules, a phone line going in two directions, a large quantity of feed, a fire cache, a weather station, a large ice box, furniture, outside lights, a siren, an indoor shower, and a number of other amenities which Buck could only shake his head at. It was luxurious compared to Lakefield Station.

Tim was in front of the office to greet them. Buck introduced Sanford and Tim introduced John Vaughan, the assistant ranger. Canyon Springs Ranger Station was not in a town. The nearest location for mail, food, mercantile and other needs was only four miles to the south, via a dirt wagon road. One of the phone lines ran to the exchange in Bristol.

At one time the Johnson County seat of Bristol was larger than Harmony. When the county went bankrupt, the land was divided between Queens County and Lake County to the west. The forest boundary was a half mile to the north of town. The Crescent River ran through Bristol

Chapter Four - 1908

as it swung around in its large loop toward the ocean. By picking up tributaries, the river became a surging broad dangerous channel of water.

The rest of the day, Tim and John showed the supervisor and deputy the station compound and discussed the places they would go the next two days. John and his wife Elisa planned supper for everyone. Elisa made regular trips to town for supplies in the wagon John and she owned. They didn't feel isolated with the road available.

Mrs. Vaughan was a pleasant woman and an excellent cook. Elisa and John had been married less than a year and it showed by their interactions and comments. Buck insisted on his usual sleeping on the ground but Sanford accepted Tim's invitation to stay in his residence.

The next day the four men rode north visiting some mining activities, small timber sales, beautiful waterfalls and moss canyons. They were almost in sight of the Crescent River, as it headed east toward Fish Cut and Harmony, before Tim suggested they turn back so there would be daylight when they got to the station.

The following day they rode south and east to some grazing areas and the only guard station on the district. The guard wasn't there but Tim showed where they were in the middle of developing a campground nearby.

Thursday morning Buck, Sanford and Tim headed west toward Ralph Hempstead's Fulton District office. The trail was in excellent condition but it still took most of the day to the town of Fulton, adjacent to the forest boundary. The district compound was on the eastern outskirts of town. It had been a large ranch headquarters when the government purchased it several years earlier. The structures were modern and the amenities were on a par with the station at Barnesville.

Buck complimented Ralph on the condition of the improvements. Ralph introduced Assistant Ranger Marshall Hubbard. Ralph was married to Priscilla and Marshall was single.

It was late, so Tim decided to stay over and ride home the next day. The five men ate another wonderful meal, prepared by Mrs. Hempstead. Ralph showed them the compound and asked if they wanted to walk to town for an after dinner drink. Buck quickly looked at Sanford for a sign. The supervisor didn't object so the five men headed toward Peacock Tavern. On the way, the deputy whispered to his boss that he was going to cut him off at one whether he liked it or not. Sanford agreed and Buck reminded him of Alice waiting on his return. He wasn't going to lie or cover for him, if asked.

The first round went down as always, in a hurry. When the second round was ordered Sanford declined and said he wasn't used to too much alcohol after meals. At this statement, Buck turned his head and could barely contain a laugh. Trying to alleviate his boss's longing for some booze, Buck brought out his harmonica and started playing tunes where everyone in the tavern joined in singing. After the third round, Buck suggested they go back because of the long day ahead. Sanford had nursed his one drink during the evening. His mind was focused elsewhere.

Tim left the next morning and the four men again toured the district for the next two days. Once an irate miner and a stockman were going to have an argument with Ralph but when they saw four men, all with rifles, there was an immediate change in attitude.

On the morning of the third day, they took the only trail north to Barnesville. It hugged the western boundary of the district, stopping at the Crescent River. The Barnesville District southern boundary picked up on the other side of a pedestrian, horse bridge. From there until the trail connected with a dirt road to the district headquarters took four hours.

They arrived at the Barnesville station with Hugh Tanner, the district ranger, meeting them at the office. He explained that Assistant Ranger Jeremy Conway was away on an emergency, so it would be the three of them on the tour. Susan Tanner's meal got Buck to thinking about the advantages of marriage but he kept it to himself. Ralph left right after dinner. He knew a family he could stay with that night, just the other side of the pedestrian bridge over the Crescent River. Sanford slept in the assistant ranger's residence and Buck slept outside.

Hugh wanted to show both men the progress made on the large timber sale from a year earlier. Half was on the Lakefield District anyway. The company planned to start up full time the latter part of April and would work until the snow flew.

Travel the next day took them near the District 3 border with District 4. It was early afternoon on Tuesday and Buck suggested instead of going all the way to Barnesville, they split and Hugh would head home while Sanford and he would continue to Lakefield. The supervisor and deputy rode east arriving at the station near supper time.

Even though the improvements were rather primitive compared to the other three district headquarters, Sanford was impressed with Buck's determination and savvy on building the place from scratch. After taking care of Titus and Gin Rummy, Buck brought out two thick, large

Chapter Four - 1908

steaks from the ice house. He figured there was enough ice to last one more year. The supervisor stared in disbelief as the ranger started a cooking fire. It would take some time for the steaks to thaw and cook so Buck showed his boss the rest of the area including the water system, ice house, phone line and hookup. They sat outside on the porch as the deputy listened to the supervisor talk about Alice. Buck wondered to himself if he had done the same with Edna after getting to know her. He hoped not.

By the time the steaks were ready, both men were famished. Sanford commented that it was the best steak he'd had since leaving the capital.

The next morning they headed east toward Morgan's Meadow. Buck told Sanford about his dream to have it built as a guard station or a campground or both. Swinging up to where the road crosses Shadow Creek, he showed his boss where he hoped someday to build a bridge and save many miles getting to the private ranches and the railroad tracks running through the district lands.

Finally, Buck took Sanford to Harris Dalton's ranch, explaining that Mr. Dalton would sell the whole thing to the Forest Service for $1,700. It would be perfect for either a guard station or even the main ranger station. The latter wouldn't work unless the bridge was built over Shadow Creek. Lakefield District was also in need of an assistant ranger and another guard. During the times that Buck told about his future plans, Sanford listened in silence. He would nod his head in agreement and now and then came out with a, "sounds logical to me."

Mr. and Mrs. Dalton were glad to see Buck and Sanford. They insisted they stay for supper.

The next morning was Thursday and Buck was ahead of schedule. He said they could go back the way they came to Lakefield for the supervisor to see more or they could ride directly south to Shadowcreek, only ten miles from Harmony. Sanford had seen enough of the forest and wanted to get back to the office. Buck knew it was something else drawing him homeward.

It was noon when they entered the Shadowcreek post office. Edna was eating a sandwich and blurted out her happiness with her mouth full. Buck laughed and leaned over the counter to give her a peck on the cheek. They waited outside at the rear of the building to let the postmistress eat in peace. Five minutes later she joined them, leaving the door open to watch for customers. After a more formal greeting, Sanford asked if she had talked to or seen Alice recently.

"Yes, Mr. Picton, I talked with Alice last Sunday over the telephone."

"Did she say anything about me or about the previous Sunday?"

"I'm not one to tell about my friends' conversations, Mr. Picton."

"Well, could you at least tell me if she mentioned my name."

"Yes, your name was mentioned."

"In a positive or negative way?"

"It wasn't negative."

"This is driving me crazy. Can't you tell me more?"

"I could but I won't. You'll have to talk to Alice."

"Buck, we're still on payroll and I have to get back to the office. Do you want to go now or stay here?"

"Well, Sanford, if you put it that way, I'd rather stay here, but Edna knows I'm still working, so we'll go together."

"Guess I know the Forest Service comes first, Buck," replied Edna.

"No, you'll always come first, Eddy."

The men made good time riding to Harmony. Lucy greeted them warmly and showed both men a table where she had laid out the urgent, the semi-urgent, the should-be-done-soon and the can-wait-awhile business matters. There was also a stack of letters, memos and reports that she had completed which were waiting for the forest supervisor's approval and signature.

Buck told her she was one in a million and Sanford concurred. Lucy thanked them and countered by asking the deputy if he wanted to help her type a stack she hadn't got to. It would be quite all right. Before answering, Buck made sure he knew she was teasing.

The two men spent the rest of the day going over the mail, reports, requests, letters, memorandums, permits and miscellaneous material. The Washington office had emphasized that no response to the public on any matter should take more than a day or two. They completed any needed correspondence before leaving the office. Buck did type three letters that were urgent. His training and willingness to master the typewriter had come in handy, as it would many times in the future.

Before the men parted, the deputy told the supervisor he would take care of the animals and unpack what remained. Buck knew his boss was champing at the bit to call Alice, so he wished him luck and said he would quiz Sanford in the morning.

That evening as Buck was eating, the phone rang. It was Kent calling from Arizona. After the usual niceties, Kent said that his parents had died soon after he arrived. He was wrapping up some business and would return to Harmony in two weeks. He asked Buck to tell Sanford. He also told Buck he had heard reports from a couple of unnamed people

Chapter Four - 1908

that the acting supervisor was drunk most of the time. He asked his deputy to please just hang on for another couple of weeks. Buck didn't deny or explain what had been going on. After all, Sanford could fall off the wagon at anytime and he was going to give Kent a list of things he wanted, as promised. The list would be preposterous but he would have some fun with it. Before hanging up, the supervisor said that he would be buying a personal motor car, once he got to Harmony. He had wanted one for a long time.

The next morning, when Buck told his boss that Kent would be back in about two weeks, Sanford groaned. He said Alice was glad he was back and had accepted his dinner invitation for that evening. Buck knew the supervisor's thoughts weren't with work that day, so he didn't mind taking over Lucy's stacks on the table. He stayed late to finish the job.

That weekend the two couples connected but in separate towns.

* * *

The time had come for Alva to take the ranger examination. Buck was one of the inspectors. There were a total of twelve candidates from all walks of life.

Buck had warned his baby brother that he wouldn't give him any special favors. He was on his own and wished him the best. He told Alva he had faith in his ability to pass. If his brother received a better grade than his, he would buy him a new saddle. Although Alva had always looked up to Buck, this offer motivated him like no other in his young life. He was determined and had spent many personal hours studying. All those months of volunteering alongside Buck and Luke were about to payoff.

The weather didn't cooperate, so the field tests took two days. The requirements had not changed much since Buck had taken them two years earlier.

After the exam, Luke and Alva brought to the office for Buck's review all the work they had accomplished during his absence. The timber volumes had been totaled to the best they could estimate and most of the district boundary survey had been completed with the appropriate signs. Buck shook his head in wonderment as to how they could have done so much in a little more than four weeks: Their answer was that they learned from a good boss and much of the work had been done during the past couple of years anyway.

Buck told them he would be joining the group as soon as Kent returned. He had a list of tasks. In the meantime, he thought they should return to Lakefield and do some maintenance work listed, plus regular fire patrol, even though it was still April. If the phone line needed repair, that should be a priority. Again, he congratulated the two on a job well done.

Sanford and Alice seemed to enjoy each other's company and Buck was continually thankful for Edna's wisdom. In a few days, he would welcome Kent home and he could go back to Lakefield. He made a list of demands for the supervisor and decided to be very serious until the end, when he would finally say, "Got you boss."

Sanford had been busy with his Washington office contacts as soon as he knew Kent would return early. He asked several friends if there was a possibility he could be considered for the Maahcooatche deputy supervisor job. He knew Kent was a wonderful person to work for, from what Buck had told him. He would be willing to take a cut in pay and swore to his close friends he had not been drinking for more than two weeks.

One newspaper article that interested Sanford and Buck was a story that the first Forest Service experiment station was established at Fort Valley, Arizona.

The Monday morning mail brought some local news. It was a proclamation that President Roosevelt had signed effective May 5, 1908. It added all or parts of 435 sections totaling about 253,000 acres to Districts 2 and 4. It was all to the west of the forest and skirted around a couple of villages. This was a huge increase for the current districts and Buck thought it should be declared a separate District 5. Sanford agreed but didn't want to make that decision. He would wait for Kent's return.

Another letter from Washington was issued to all forest officers. Describing timber marking procedures, the instructions suggested officers doing the marking try to visualize how the area would look after logging, making it good enough to demonstrate a fine example of forestry to the general public. It further recommended leaving all young growth below merchantable size and at least one-third of the present stand of merchantable timber. It was to plan for a second cut in 30 to 40 years. This letter was one of the first approaches to marking rules.

The days before Kent's arrival didn't go quickly enough for Buck. On the other hand, they went too fast for Sanford. The supervisor wired he would be on the Tuesday morning train. Buck said he would meet the train and carry the luggage to his boss's house. He planned to remove his belongings and stay at the boarding house until he could leave town.

When the time came, the acting deputy had made sure all the corre-

Chapter Four - 1908

spondence was caught up and the public's problems were largely taken care of. Buck waited fifteen minutes at the depot till the train pulled in. They would walk to the house.

Kent waved and shook hands with the deputy. They were both glad to be back to normal, even though the acting supervisor had been a perfect gentleman since the show-me trip. As they trudged to Kent's house, Buck gave a blow by blow account of the past few weeks. Then the ranger asked the supervisor how he felt and Kent quickly said he was fine and ready to return to work.

Buck said, "Well, my friend, do you remember what you told me before you left?"

Kent replied, "I said so many things, I don't know which you mean."

"It was about the list of demands I could present to you if I baby sat Sanford successfully."

"Oh no! You caught me with my back against the wall, Buck, but I'm a man of my word. Let's see the list."

"It isn't long. The first one is I've run up a bar bill of more than $25 to pay for the phantom drinks that Sanford asked for at the various saloons. The second one is to address the local high school graduating class. I was scheduled to do it next week but now I bow to your infinite wisdom and superiority."

Kent let out a howl. "That last part is a bunch of bull and you know it."

Buck went on, "Edna and I feel you need a female companion. Therefore, Edna plans to set you up with an acquaintance of hers who lives in Fish Cut. We'll tell you when and where.

"Finally, you mentioned buying a personal automobile. I've always wanted to drive one of those contraptions. I want your vehicle to be available whenever you aren't using it. There you have it. No problem. Right?"

Kent was beside himself. He yelled, "That's blackmail, Buck, all of it. I didn't think you'd ever be that lowdown. I still don't believe you—" he didn't finish the sentence.

Buck couldn't hold out any longer. He burst out laughing and rolled on the couch. Tears streamed down his face and he couldn't talk for a few seconds.

"You mean to tell me this was all a joke and you don't mean any of it?"

"Of course, Kent. How could you ever think I'd be that rotten? Got you boss."

Now the supervisor started to laugh. It took awhile to regain their composure.

As they walked to the office, Kent did tell Buck that he would pay for the bar bill. Buck protested. They settled on a 50-50 split. Buck would make the graduation address, since he had already written the speech. The supervisor planned to buy the auto sometime next year. He had heard Henry Ford was planning to build a low priced mass produced model. Buck could drive it at times. As for Edna's acquaintance, Kent felt he could choose his own if left alone, but wouldn't mind if Buck and Edna helped out sometimes, making it a foursome just as they had with Sanford and Alice. He knew Alice but not well. She seemed like a fine lady. Buck agreed.

Lucy greeted her boss with a big it's wonderful to see you welcome. Buck introduced Kent to Sanford and the three closed the door for a meeting.

Sanford was scheduled to return to Washington in two days on Thursday morning. He told Buck he was going to ask Alice to marry him that evening. The ranger wished him well.

The next morning, Kent, Buck and Lucy were at the office but there was no Sanford. At mid-morning Buck said he would check at the hotel room. He was worried. There was no answer when he knocked and no acting supervisor when the clerk opened the door. Now what, thought Buck. As he left the building a familiar figure staggered down the street. "Oh no!" the ranger said out loud. He hurried to Sanford and helped him to his room. He was a mess. From what Buck could gather, Alice had turned Sanford down mainly because she wouldn't even consider moving to Washington, DC. She did tell Sanford that she was very flattered to be asked but didn't know at that time if she loved him, anyway. He had kissed her goodnight and goodbye a few hours ago and then purchased a couple of bottles and went on a real bender. He didn't remember where he had been or what he had done. Buck called Kent from the lobby office and told him he would stay with Sanford all day and night, if necessary, to make sure he got on the train.

It was one of the longest 24 hours Buck had ever experienced. Sanford was sober by supper time, so they ate together at Golden's and Buck stayed in the same hotel room that night. After packing his luggage, the acting supervisor asked Buck to sit down. He had something to say.

"Buck, you are undoubtedly the most loyal and best friend a person ever had. I recognize what you've done for me since I arrived. I've put you through hell and I apologize. The only way I can repay you is through some of my connections in Washington. I plan to talk to my senator uncle and ask him to try to come up with some special money for

Chapter Four - 1908

you to build your bridge, buy your guard station, build your campground and hire some help. You deserve it. I don't know if it's possible, but I'll try. I know that the powers in the Forest Service do not bend to political pressure and I admire them for that. I'll approach the matter as being worthwhile and beneficial to the public and the Service. What do you say?"

"I'm overwhelmed, Sanford. I never realized you felt that way. To be honest, I did enjoy your company when you were sober. You have a lot of savvy and can be an asset to the Service, if you stay on. As a word of personal advice, I wouldn't give up on Alice. You can always write and if she doesn't want to move, find out what would be acceptable and take it from there. We better get to the depot."

The two men walked and talked as they each carried a suitcase. They shook hands as the boarding began and Sanford was gone. Alice had not been there to see him off. Buck didn't think she would be.

* * *

Kent requested that Buck remain till the first of July. The ranger reluctantly agreed. The supervisor spent most of his time grappling with the added forest acreage. He didn't see how Ralph Hempstead and Hugh Tanner could take on half again as much land as they already had.

In the meantime Buck and Alva received great news. Alva passed the Forest Service ranger's exam just three points below his older brother. He wouldn't acquire a new saddle just yet.

Buck requested a meeting in Kent's office with Alva present. He outlined his belief the forest needed to make the added acreage into a new District 5. He also believed his brother would make an outstanding ranger on the new district, emphasizing he was talking as an acting deputy supervisor instead of a relative. Kent agreed completely but said there was no salary money, let alone funds for an office. The forest-wide request for fiscal year 1909 starting July 1, 1908, had been sent in before the proclamation was issued. Buck said he would write a justification for an added district, including salaries and rental or building funds for the initial year. It took him two days to organize and write. It was a masterpiece of logic, reasoning, public and Forest Service benefits and the dire results if it was turned down. Kent happily signed the analysis.

The mail usually had some copies of press releases or directives from Washington. One dated May 23 caught the eye of Buck and Kent as they reviewed the literature. A new law known as the "25% Law," provided funds for counties containing national forest lands. Twenty-five percent of all revenue from timber sales, grazing permits, recreation fees and several other sources was to be returned to the counties. These revenues were in lieu of taxes that the counties would have received if the federal government had not taken over administration of the public land. The funds were to be used for public schools and roads. Previously, the funds had been at 10% of all revenue.

It was approaching the time when Buck would return to Lakefield District. Sanford had been gone for several weeks and the ranger had forgotten the acting supervisor's speech during his last hour at Harmony.

One morning after the mail was distributed, there came a loud exclamation from Kent's room. He yelled for Buck to come immediately. The news was unbelievable. The new fiscal year fund request had been approved but added to it was a supplemental listing for money to buy a guard station, build a bridge, construct a campground and salaries for an additional guard and assistant ranger—all for District 3. It totaled several thousand dollars but needed to be spent by June 30, 1909. Both men stared at the document for several seconds, expecting it to disintegrate before their eyes.

Buck said, "I can't believe it. Sanford did what he said he would try to do. I'd forgotten all about it. In fact, I didn't give it a second thought. It sure is great news for the forest."

Kent replied, "You've done it again, my friend. Your attentiveness and patience resulted in this windfall. Congratulations. What do you want to do first?"

"First, I want to buy the Dalton Ranch. Harris wants to sell as soon as possible. Second, we need to build a bridge across Shadow Creek close to our new guard station and Morgan's Meadow. Finally, I would like to build an improved campground at the Meadows. With the road from Shadowcreek and the trail across the bridge, it would be accessible and a perfect spot with a corral, water, trees and flat areas for campsites."

"What about this money for a guard and assistant ranger?"

"Well, Kent, it's money you can use for any salary you decide. I've been thinking that for the good of the forest, I would forego any additional Lakefield District men and ask you to use it starting July 1st to pay for Alva's district ranger salary plus a guard to help him out."

"You'd sacrifice added men for next year?"

"Yea! Luke and I have been doing the work for the past three years, so one more won't make much difference. The only request is that we will probably need an extra man to install the bridge."

Chapter Four - 1908

When Kent told Alva the good news and how it came about, tears rolled down the latter's cheeks. He vowed to try and be as good a ranger as his brother. He would be on the payroll in a few days and along with his new guard, would have the responsibility of administering thousands of acres of public land. He was humbled and impatient at the same time. What a brother!

As expected, Harris Dalton sold his ranch for $1,700. Mr. and Mrs. Dalton and Buck were happy and toasted each other with drinks and an excellent meal prepared by Mae Dalton. The ranger assured them they could take their time moving out. He intended to concentrate on the bridge before the ranch. It was not to be!

* * *

The July 1, 1908, edition of the *Use Book* was issued. It had increased from 142 pages in 1905 to a whopping 341 pages. Being almost an inch thick, it wasn't practical to place a copy in the ranger's shirt pocket. It would be stuffed into the saddlebag.

For several years Buck had issued class A and class B timber sales. Since type A was considered a ranger sale not over $50 value for dead or living timber, there was no delay after the applicant paid the money and agreed to the terms of the sale. Class B supervisors' sales were not over $100 in value. Although Buck did the same for both types, he couldn't sign the contract. In this case he would forward it for Kent's signature. The public realized it was a good deal to be able to cut and remove timber from the forest, so more and more sales took Buck's time. Final inspections took many extra hours if there was a trespass or the purchaser hadn't followed the stipulations.

Class C sales approved by the forester were for more than $100 in value. These had to be advertised and a bond required for those over $3,000.

All uses of national forest lands and resources, except those which related to timber and grazing, were known as "special uses." Applications for special use permits were sent to the forest supervisor. Depending on the type, the permit was signed by either the supervisor or the forester. However, the ranger had to do the administration once it was in effect. Buck had to deal with residences, farms, pastures, stores, mills, railroads, camps, summer resorts, tanks, dams, power lines and several other types. The charge for permits was based chiefly on the value of that which was actually furnished to the permittee by the Forest Service, not directly on the profits of business which was to be carried on.

There were also free use permits. These included cemeteries, churches, municipal water plants, schools, drift fences, corrals, roads and trails, telephone lines, irrigation conduits and others.

The last half of 1908 Buck and Luke were busy with homestead claims. They were both fighting fires at various locations and times on the district. None got out of hand, due to their diligence and active suppression action, including hiring several local folks in the forest boundaries.

During September, Kent sent each of the now five rangers a circular letter of instruction with a sample "Summary of Ranger's Service Report," form 347. The following activities were listed for charging ranger's time:

Timber Sales	Trespass
Fire Patrol	Forest Plantings
Survey Administrative Sites	Special Uses
Claims	June 11 Examinations
Boundary Survey	Grazing
Free Use	Improvements
Fighting Fire	Office Work
General Administration (Misc.)	Sundays/Holidays

This added another layer of work to the ranger's time.

Winter was coming and Buck acknowledged his bridge would have to wait until spring. Luke was off the rolls and the ranger decided to divide his time between Lakefield Station and the Dalton Ranch, renamed Dalton Guard Station. He had plenty to do at Dalton such as painting, repairing fences, replacing building lumber, putting up signs, bringing in feed and making the office and residence habitable with furniture, ice box, pots and pans. Edna visited quite often and fixed the place with curtains, bedding and many other niceties which only she could do.

The supervisor had received instructions concerning the installation of weather stations. Data was required from three months at high stations and year long at lower elevations. The Government was to pay up to $7 a month for local, cooperative observers. Buck decided to build a weather station at Dalton Guard Station and Lakefield Station. He would save the monthly payments.

* * *

In 1906, the Western forests had been divided into three administrative districts—the Northern Rockies, the Southern Rockies and the Pacific Coast. These offices were primarily for inspection work. Earlier in

Chapter Four - 1908 139

the year, an inspector was sent to the Maahcooatche. He could find no fiscal improprieties or any major problems. His report reflected that Kent was an excellent supervisor who gathered outstanding people for district rangers.

A November 5th letter signed by Gifford Pinchot established six district offices, effective November 1st. Each district was divided into national forests, managed by a district forester. Mail and personal contacts would now flow between individual forests and the district office they were assigned to. Forests would not communicate directly with the Washington office.

Near the end of the year the forest received another circular proposing four styles of Forest Service uniforms with specifications and price lists for each type. The fabric would be made by the American Woolen Mills. The uniforms would be made by Fechheimer Brothers Company. Stetson would furnish the hats, which would have a four inch crown and three inch stiff brim, for $3.45. Uniforms would not be made mandatory but should be worn to create a good appearance to the public. Kent decided the forest officers would vote on the style preferred.

Edna and Buck spent New Year's Eve celebrating at the community dance. They were the most graceful couple on the floor. It had been Buck's hardest year yet as ranger, except it hadn't been the ranger part that was hard. They toasted in the new year and hugged and kissed as if no one else was present or mattered.

1909

The Call

We who alone are wont to ride
Among the pines at eventide,
And climb to where some jutting crest
Gigantic looks toward the west,
There at the sunset hour to seek
O'er wide-flung realms of crag and peak
And canyons, black with mystery—
Gold islands in a shadow sea
Where silent tides of purple shade
Engulf red shores that glow and fade—
   Ah, we have heard the Voice that calls,
      That magic Voice which has no sound:
   From out the dusking night it falls,
   From canyon's depth and granite walls,
      And awe has compassed us around.

And lone the trails we ride that run
Where canyon shades shut out the sun:
Rock-gated is the op'ning pass
Whence bursts the mountain's awesome mass,
Where, far above the proudest height,
A searching eagle hangs in flight
And, ever soaring, wheeling, throws
A circling shadow on the snows:
And darkling is the forest shade
When camp by dusky stream is made—
   Ah, then the hobbles' clank we hear,
      When packs are off, and saddles thrown,
   And, breathing round the campfire's cheer,
   Again the silent Voice draws near—
      The mountains, calling to their own!

Scott Leavitt

CHAPTER FIVE

1909

Buck was at Lakefield station restocking the ice house. The first ice had lasted two years and was responsible for many happy moments with people savoring delicious steaks, ribs and other tasty morsels. In fact, word of the ranger's supply of meat in the middle of the forest traveled far and wide, so that he often had unexpected visitors ride or walk into the station hoping to be invited for a meal. If they behaved, Buck never failed to treat them. If called away for an emergency, he would direct the guests to take their time but to clean up afterwards. He didn't want bears or cougars roaming around as the result of loose meat scraps. Most of the visitors had dogs, anyway. Buck never accepted any money or gifts. He did give everyone a low key lecture on keeping campgrounds and trails free of litter.

His meat inventory consisted mainly of beef with a little lamb thrown in. During the fall he stocked the ice house with venison.

It was a beautiful crisp, sunny day and the last haul for the toboggan from Green Lake to the nearly filled ice house was in progress. Suddenly a strange sound floated through the forest. Buck stopped and listened. It was like a crying wail at five to six second intervals. He was sure it came from the north near the lake. Not knowing what to expect, he jumped on Titus and rode cross country all the time yelling and then stopping to hear if he was going in the right direction. The sound became louder. Coming around a large rock, Buck saw a middle aged man lying on the ground with his right pant leg torn and blood coming from a wound. It was apparent his leg was broken and he was in agony. Buck dismounted, took his scarf and tied a tourniquet around the man's thigh, all the time assuring him that he would eventually be all right. Knowing the man couldn't walk, the ranger said he had to get his toboggan and would return in about 20 minutes. This panicked the stranger but Buck took off immediately and was back with Titus and the toboggan within the stated time. The bleeding had stopped but the man was in shock and didn't comprehend much of what the ranger was saying. Buck laid him on the sled and strapped him in. It was a rough trip to Lakefield.

The phone was down, so Buck retrieved Stub and told the man what he had to do. There was no other way but to mount the mule and ride to Shadowcreek.

The ranger had no experience with setting leg breaks but he did tie a temporary splint down the right leg to keep it immobile and lessen the pain. In the meantime the man said his name was Leland Peel and he had broken his leg by stepping on what he thought was solid snow. It wasn't. His leg had twisted and the clothes had torn on a jagged rock outcropping from the hole he fell into. He was headed for town and knew there was a ranger station in the vicinity. The irony was that he would get to both destinations but not as a hiker. When Buck asked why he hadn't worn snowshoes, Leland said he had hiked all over the West in all types of weather and didn't anticipate any problems. This was his first injury. Making sure they had plenty of water, Buck and Leland rode for Shadowcreek. With several stops and a slow gait it took six hours to reach the front door of Doc Cleary's house.

Buck stayed with the patient while the doctor set the leg. It was a clean break but Leland had suffered a great deal. He fell asleep in the doctor's operating room from pure exhaustion. The few personal belongings Buck had retrieved from his clothing were given to the doctor. It was too late to return to Lakefield. Titus and Stub were stabled at Ike's Livery and Buck walked to Edna's house. She wasn't home so he sat down and waited. As soon as he saw her, with two large baskets loaded with packages, Buck sprang up to help. Edna hadn't expected such a wonderful surprise and she expressed her delight with a warm kiss planted on the ranger's whiskery face.

After closing the post office, Edna had gone shopping at the mercantile and obtained some gimp[17] for trimming her furniture and several yards of cambric and chambray.[18] She had also purchased eggs, flour, sugar, butterine,[19] canned peas, corn and Carnation evaporated milk. Edna was an excellent seamstress. One time Buck had watched as she made her fingers and hands move in perfect coordination as her feet used the treadle on the sewing machine.

Buck explained why he had to come to town. He called the Pilot Hotel and asked Faye to save him a room. Eddy and he ate in that evening.

* * *

Chapter Five - 1909

All winter Buck thought about how the suspension bridge would be built over Shadow Creek. He was anxious to get the job started and completed before all the other summertime duties consumed his time. In fact, it almost became an obsession. He had made a drawing and determined the amount of material for the cable, transverse members, planks, bolts, towers at each end, vertical rods, washers, nuts, rope and miscellaneous items. He had come up with a plan to go back and forth on a temporary bridge on hands and knees. He also listed the tools needed, including a forge, drills and two tall ladders. The holes would be hand driven and the forge using fir bark would sharpen the steel. It was a three man job, so along with Luke, starting May 1st he would have to borrow from some other district.

During April, Buck had worked on the trail from Dalton Guard Station to Shadow Creek, a distance of about five miles. He expanded it into a primitive wagon road in order to move all the equipment to the site.

The first thing Buck did when Luke came on board for the usual six months was to have him meet at Ike's Livery in Shadowcreek. He had ordered the material a couple of months earlier and had it delivered to the mercantile store. The ranger had approval for the bridge design and location from Kent and had also received permission from Tim Westgate to use John Vaughan, the assistant ranger on the Canyon Springs District. Buck had been impressed with John's work and attitude during the show-me trip the previous year and was extremely happy to have him as his third person.

Luke and Buck took most of the day to load up and inventory the necessary materials, tools, food and essentials for the project. They rented a wagon from Elroy. John arrived that afternoon and the three men headed north to stay overnight at the district's new Dalton Guard Station.

By the next evening all the material was laid out at the creek's eastside edge. A flat spot for one of the towers was prepared. It was necessary to hand drill holes two inches in diameter in the rock for the anchor pin which would then have the seven-eighths galvanized cable looped through the eye and clamped back onto the cable. The towers were constructed as a unit. They had to be high enough so that a horse and rider could ride beneath the lowest cross arms of the wooden structure. Buck had figured the length of each cable to be a little more than twice the distance between towers. The slowest work was the hand drilling into solid rock.

To save the long horse trip around the road and down the westside trail, Buck had figured to construct a simple three foot wide bridge so cables and other material could be inched across from the eastside to the westside. He had earlier tested a 30 foot, three inch diameter hollow iron pipe. Each pipe was in five, six foot threaded pieces. The tricky part was to place the two 30 foot pipe rails next to the bridge site. Each rail was raised individually to a vertical position. The bottom end was braced against the wagon. The top end was tied with three ropes fanned out so that each man slowly lowered his rope and dropped the rail to the opposite side. The two rails were rolled to about three feet apart and the four foot board sections dropped into place. By crawling, the men could now move from one side to the other.

The holes and tower were completed on the westside. It was time to attach the various sized rods at four foot intervals on the cable section that hung between the towers. These would hang down to eventually hold up the bridge planks. There were nine rods, drilled with holes at the top to be able to clamp to the cable. The bottoms were flattened and threaded for washers and a nut to hold the individual six foot long, six inch deep and four inch wide transverse members. After the cable was anchored at both ends and crossed over the towers, they placed the catenary cable so the lower end of the rods were at the proper level of the planks.

The next step was to tie the wooden transverse members between the two cables. Planks were extended in mid air so that Buck could crawl out with a previously drilled member, insert the rod end, place a bolt through the hole and tighten it with washers and a nut. The other two men stood on the two by twelve planks for a counterbalance. This procedure continued until all nine members were attached and the rods were hanging loosely. The final job was to attach the planks to the members. It had taken the three men two weeks from start to finish. That evening back at the (now easily accessible) Dalton Guard Station there was a party with some tall tales joined by John Barleycorn.

During the winter Buck had also laid out plans for a Morgan's Meadow Campground. For the next several weeks, he and Luke worked to complete the project. They could take the lumber, tools, hardware and equipment as far as the bridge in a wagon, but although it was a short distance, all of it had to be moved on the pack train from the bridge to the campground site.

Tables and benches were built, privies were constructed, holes were dug, fire pits were installed, areas at each campsite were leveled for tents, a small corral was erected and posters were placed for rules and

Chapter Five - 1909

regulations. Luke said he thought it was the most beautiful campground on the forest.

When Buck next reported to Kent, the supervisor praised his accomplishments and his use of the money before July 1st.

Buck had one month before Kent requested each ranger report for their annual meeting. In the meantime he had asked Luke to get the phone line in working condition and make an up-to-date property inventory, since an inspector was expected sometime during the year and accountability was one of their targets. He, in turn, would check out some of the mining claims on the district.

The 1872 Mining Law was the basis for allowable mining on national forest lands. Claims were made through the county recorder. The maximum boundaries of a placer claim was 660 feet by 1,320 feet; or 10 by 20 chains. The maximum boundaries of a hard rock or lode claim was 300 feet on each side of the vein by 1,500 feet along the vein. Both these measurement came to approximately 20 acres per claim. There was a dollar amount assessment work to be performed on each claim. It required $500 for the first year and $100 each year thereafter. The number of claims an individual could have was not limited, so long as mining was performed and assessment of improvements was done. The claimant had to be a United States citizen and a corporation could be considered a citizen. Associations of mining claims were created to where eight individuals could each make a claim. Based on a single discovery, abandonment of a claim meant someone else could claim the area.

Buck had a checklist of three items to look at. He wanted to make sure the claimants were not just living on the claim with no mining activity taking place. He wanted to make sure the cutting of timber used to support the sides or back of a working lode mine was within the area of the claim. The only exception to this would be if the timber had burned over. Then an adjacent area with timber could be used. Timber could also be used for a cabin and privy. The third item to check was a requirement by the state for the miner to install a specific sized post at each corner of the claim as well as the discovery site.

A long list of minerals and rocks were allowed to be claimed under the mining law. Among these were building stone, china or fire clay, coal, limestone, oil, salt, slate, gold, silver, cinnabar, copper, tin and lead. On lode claims, the *Use Book* stated the forest officer should examine the dimensions of the outcropping of the vein or lode on the ground. He should search for discovery shafts, pits or tunnels which the owner had made in an effort to find or disclose the vein and ore. He should take measurements and samples of each important grade of ore found. These samples should be wrapped, labeled and stored for further analysis. On placer claims, the forest officer should make several pannings in each excavation. His report would verify or dispute the nature of the deposit claimed.

Buck knew all this and was extremely glad he didn't have any lode claims on the district. They were all placer but he wasn't going to spend time panning unless there was something that raised his suspicions.

Buck had no trouble with the first dozen or so miners. He was friendly and so were they. A couple of times the claimants had jumped to conclusions about the ranger's presence. They thought he was there to either close them down or issue a citation. Buck explained he did not have the authority to do so. He was there only to check on his list of three items. On several occasions he instructed the miners to install the posts.

On one he found an old man living in a log cabin, eating from a garden but doing no mining. The man explained he knew it was wrong but his health wouldn't allow him to do the necessary physical activities. Upon further inquiry Buck learned he had relatives in Barnesville but they didn't know his plight. The ranger asked the old miner, Sidney Porter, if he would be willing to gather up his belongings and ride Stub to town. With tears in his eyes he answered affirmatively because he knew his condition wouldn't allow him to live alone much longer. He emphasized to Buck that it was breaking his heart to have to leave his beautiful forest home. Buck placed his arm around Sidney's shoulder and said he understood his feelings and that if he had problems he should talk to Hugh Tanner, district ranger at Barnesville. This seemed to somewhat cheer up the old man. After gathering his meager belonging, Buck, Sidney, Titus and Stub headed for town.

Sidney's relatives were surprised but happy to see him. They thanked Buck and the ranger rode to visit Hugh. It was just a few days before the meeting in Harmony so the two rangers traded information on the current timber sale, employees, funds, projects and other common interests.

Buck had one more mining claim to investigate. A miner he had checked previously reported some gunfire a few days earlier. It could have been hunters or someone shooting at a stationary target. He wasn't too concerned and had ridden for some time when from out behind a tree a man with a long beard jumped and held his gun directly at the ranger's heart.

Chapter Five - 1909

Buck stopped and the man spoke, "Mister, you're trespassing on my property so turn around and get going or I'll place a bullet where it won't do you any good."

Buck calmly responded, "Why are you upset, sir. I'm just riding through these woods enjoying a sunny warm day, minding my own business."

"I don't know you and I don't trust strangers."

"My name is Buck Stonewall and I'm the ranger on this district."

"Well, I don't trust rangers, either."

"Why don't you tell me your name and then we can talk easier."

"Name is Jonah. That's all you need to know."

Buck dismounted and walked toward the gun.

"Don't try any funny stuff, Mr. Ranger."

"Don't intend to, Jonah. Why don't you put that gun down and I'll stop right here."

"I'll point it toward the ground. Now what do you want?"

"I had a report of gunfire in the area a few days ago. Do you know anything about it?"

"I might and then again I might not."

"Well, just suppose you know something about it. Why not tell me and I'll leave it alone."

"You wouldn't believe me if I told you."

"Give me a chance. I'm not here to play questions and answers, Jonah. As the ranger, I'm here to check your mining claim." Buck listed the three things he wanted to look at. "Let's go to your living quarters and discuss this sensibly."

"Well, I don't want more trouble, Mr. Ranger, so follow me. It's only a few feet behind that stand of trees."

Buck walked alongside Titus. He noticed a freshly made mound of dirt like the one in a cemetery off to the side. He said nothing. Jonah was a large man with a fully whiskered face and long hair. He walked with a limp and seemed unduly nervous. As they rounded some trees Buck noticed an old log cabin about 10 by 14 feet. There were two chairs on the front porch and Jonah offered one to the ranger.

"We can sit here and talk, Mr. Ranger," said the miner.

"Fine with me, Jonah. I'd like to see where you're mining and check your corner posts before I leave."

"I'm placer mining in Lookingglass Creek about 50 yards from here."

"How long have you been mining here?"

"About a year. I came across this unoccupied cabin, checked it out and found I could file a claim after a search showed the previous owner had abandoned the site." By this time Jonah had placed his rifle against the outside wall of the cabin.

"That's interesting, Jonah. Yours is the first mining claim I've come across on the district that was abandoned. I noticed a mound of fresh dirt as we came in. Would you tell me what that's for?" Buck said this last part casually as an afterthought. The miner didn't say anything but looked at the ranger very intently. Finally he said, "You know, Mr. Ranger, I'm usually a peaceful man who just wants to be left alone. I stopped you earlier because I didn't know who you were or what you wanted. I've had a hard time the last few days. Haven't been able to sleep or do much work. My old mule is out back in a makeshift corral. You seem like an honest and just man so I'm going to tell you a story. It's all true. You can say what you will after I'm finished."

Jonah went on to describe that a few days ago another man had stopped and wanted to know what he was doing trespassing in his cabin. He had told him he was living there and mining, explaining he had a valid registered claim. The other man, calling himself Black Pete said it was his claim and that Jonah should leave immediately. He saw the man had a holstered revolver and was quite wild-eyed. So he gathered his few belongings and walked to his mule, Lily. Realizing he had forgotten his rifle he headed back to the cabin. Black Pete had left for a few moments so he retrieved his rifle and walked out the door. At that moment he heard a shot and felt a hot pain flow through his left calf. Falling down he looked to where the noise came from and saw Black Pete taking aim from about 50 feet away. Another shot rang out and hit the front porch beam near where he was lying. Jonah explained he thought he would be killed so still holding onto his rifle he took careful aim and fired. Black Pete died almost immediately. He had left the body in the same location for the whole day. He supposed he was in shock. The next day he buried it.

"And that's the whole story, Mr. Ranger. I didn't want to kill anyone but I had to do something in self defense."

"Jonah, I believe your story. I had a similar run-in with Black Pete four years ago. He was a short tempered, violent man who lived a precarious life."

Buck went on to say that Jonah had two choices regarding the killing. He could stay at his claim and wait for the sheriff, or both of them could ride to the sheriff and explain the situation. He told Jonah that he believed the sheriff would realize he had turned himself in with a logical explanation. Along with his own verification and the sheriff not having to spend a couple of days in the woods, the odds were in his favor if he made the latter decision. He asked if he had the revolver. Jonah brought it out from behind some clothes. His left calf was still in need of medical attention so Buck, Jonah, Titus, Stub and Lily aimed directly for Crescent Ranch, the closest first aid station.

Sarah Parley was home alone and immediately took Jonah in and cleaned the wound, applied some medication and wrapped the leg. According to the miner, it felt much better. Sarah guessed that gangrene hadn't set in yet. It had been a long day and she offered the whole party a place to eat and stay overnight. Luke didn't answer the phone at Lakefield. The boys were on a camping trip with friends.

Buck explained to Jonah that they would leave early for Doc Cleary's. Then they would ride to Harmony to the sheriff's office and tell the story. If all went well, the miner could follow him back to Lakefield station and from there to his mining claim. The next morning the two men thanked Sarah for her hospitality and rode east along the Crescent River road to Shadowcreek.

Doc Cleary said Sarah had done a terrific job on the wound and Jonah shouldn't have any trouble with infection or walking after a few days. They left for the sheriff's office without stopping at the post office.

When Sheriff Miller saw Buck come in with another man, he assumed Jonah was a criminal like all the others the ranger had provided. He said, "Not again, Buck. I just emptied the jail yesterday."

"No, no, sheriff, it's not what you think." He went on to explain his inspections of the mining claims on the district and a blow by blow account of his meeting with Jonah. He let the miner take over and tell the exact same story as Buck had listened to a day earlier. Buck said he had known Black Pete in an earlier encounter and agreed that his actions were in character. He handed the sheriff the revolver and asked if they could leave. Sheriff Miller had to know Jonah's last name and if it was the same as registered at the county office. The miner said it was and his last name was Miller. At this, the sheriff almost swallowed his chaw of tobacco.

"Small world, isn't it, sheriff?" exclaimed Buck as they left through the front door. They walked to the Maahcooatche's headquarters. Ranger Stonewall needed to check with Kent for any last minute changes with the coming meeting.

The supervisor was glad to see Buck. He wanted to show him something that had arrived two days earlier. It was apparent that Kent was excited and sought to share his pleasure. Leaving by the rear door, the three men walked up to a Ford Model T automobile. The supervisor enthusiastically described it. It was a 1909 touring body type, weighing 1,200 pounds. It had set him back $850. He could have had either red or Brewster green.[20] He chose the dark green. It had metal running boards with a series of parallel ribs running lengthwise, full black leather interior with 30 by 3 inch tires in front and 30 by 3 1/2 inch tires in the rear. There was a three pedal system marked "C" for clutch, "R" for reverse and "B" for brake. The steering column assembly had brass plated spark[21] and throttle levers with hard rubber knobs. The starting hand crank handle was also hard rubber. Kent said he had received one of the first ones since production began in October of 1908.

"There it is, Buck. Just like I said. Isn't it a beaut?"

"Well, I'll be," exclaimed the ranger. "Yes, Kent you got it, but how do you expect to ride to Lakefield?"

"Don't be funny. This isn't government owned."

"Remember, you said I could drive it sometimes."

"Yep! And, I'll keep that promise but you have to have driving lessons first. I bet you don't even know how to start it."

"You're right. I'm depending on your goodwill to teach me," Buck responded with a laugh.

"I'm learning, too. Soon I'll know something that you don't and that'll be a change," Kent said with another laugh.

All this time Jonah was walking around the car, feeling the fenders, seats and tires. He said it was quite a contraption.

The meeting was in three days and a number of important items would be discussed at that time according to the supervisor.

"How's the fixins at the Dalton Guard Station? Is it suitable for eating and sleeping five rangers and a supervisor?" asked Kent.

Buck responded by saying it was but he didn't have the money to feed everyone. The supervisor told him to figure a three day meeting for six people and to buy all the food necessary and charge it to the Forest Service, including the transportation to get it there.

If they planned to have the meeting at Dalton's, Buck said they should probably contact Hugh at Barnesville and Alva at the District 5 headquarters at Long Tom Ranger Station, so they wouldn't go out of their way by coming to Harmony first.

Long Tom was a village about the same size as Fish Cut. It was just inside the Lake County boundary, adjacent to Queens County. Ingot, a large town west of Long Tom was the county seat. Alva had originally

Chapter Five - 1909

thought of working in an office at Ingot but decided it was too far from the district boundary. Long Tom was within the forest and Alva found a building to rent for a ranger station. His new fire guard would be at the same station temporarily. A trail connected Long Tom with Ingot. From there, a good wagon road went north, south and west.

Kent said he would contact both rangers. He figured to drive directly to Dalton's with Tim and Ralph.

"I thought you said it wasn't a government vehicle," said Buck.

"It isn't but I can use it if I pay for it," retorted Kent.

"Yea! But what happens if you break down on the way. Are you going to take yourself off the payroll?"

"Do I sense a bit of envy for not having an auto to drive?" the supervisor asked.

"No, sir, I wouldn't sell Titus for three Model T's. Anyway, Kent, I've got to get Jonah back to his claim and buy groceries for the gang. I'm looking forward to the meeting. See you later."

Buck and Jonah mounted their animals and left. A quick stop at the post office filled Edna with the latest news and the fact that six men were going to take over the Dalton Guard Station for three days. Buck said he would return the next day to buy the food and the supplies necessary for the meeting. That evening Buck and Jonah had steaks at Lakefield.

The next morning the two men traveled west for a few miles and parted after the miner assured the ranger he could find his way. Buck had made a friend for life.

Returning to Lakefield, Buck expected to continue right on through to Shadowcreek. Luke had recently arrived from the north and was preparing a trip to Tip Top Mountain. They sat on the front porch and talked about their last few days. Luke had toured the entire district and installed trail signs at intersections. He had measured distances by using the map. The telephone line was repaired and all the government property was inventoried at Lakefield, Crescent, Dalton's and various campgrounds. He gave a copy to Buck, who in turn explained what he had been doing checking mining claims. He also said that the five rangers and supervisor would be at Dalton's for three days and Luke was welcome if his travels took him that way.

It was afternoon when Buck left for Shadowcreek. He made a mental note of what and how much food would be needed for six hungry men during three days. He gave a written list to Calvin Nibbs, the owner of the Shadowcreek Mercantile and asked if the list could be filled by the next morning.

"Absolutely," said Calvin, a small baldish man who wore glasses and talked in stops and starts, but was always accommodating and in good humor.

Buck left the animals and asked Elroy for rental of a wagon for the next five days, charged to the Forest Service.

Edna and Buck spent the evening at her house discussing the past couple of months and the upcoming meeting. When Buck told her the list of food he had ordered, she asked about a number of items he had forgotten.

"Gosh! I sure didn't think about all that stuff, Eddy."

"That's because you don't really cook, Buck. It's either meat from your ice house or bacon and beans with Titus. I think one of the questions that should be answered by every potential ranger in the Forest Service is, 'Do you have a cast iron stomach?' If the answer is no, then they should do something else."

Buck looked at her intently to see if she was serious or not. She burst out laughing from his look.

It was late when the ranger entered his room at the Pilot Hotel.

With the added list of food, Buck was delayed in leaving town for Dalton's. Arriving at the guard station he unloaded, took care of the animals and made sure there was a bed of some kind for everyone. He would sleep in the barn or outside as he preferred. The meeting would take place in the living room. All was in readiness.

Hugh arrived first. Alva came fifteen minutes later. Buck was almost ready to go search for the other three when they heard a chug, chug, chug coming up the road. All three were waving arms and yelling as they pulled into the front yard. Buck, Hugh and Alva bent over with laughter. They wouldn't have believed it if they hadn't seen it with their own eyes. With a loud bang the engine coughed and died.

Kent asked Buck to give them all a quick tour of the new station. A half hour later Buck received rave reviews on his variety and choice of sandwiches for lunch along with fruit and cookies. His mind flashed back to Edna who had saved him from humiliation with the addition of what he had forgotten.

After lunch they all settled into the front room and the supervisor brought out a large packet of papers and some charts, maps and circulars.

"Well, men, we have a great deal to cover in the next few days but first I want to praise you all on a job well done since our last meeting a couple of years ago. I am the envy of all the other supervisors I know.

Chapter Five - 1909

Don't get too comfortable, however, because we have more responsibilities added this year."

He went on listing the accomplishments the forest had made and the status of the past marking of boundaries, Homestead Act claims, mining claims, property inventories, timber volumes, timber sales, campground construction, trail improvement, trail signs, fire preparedness and suppression, water measurements, wildlife reports, road construction, bridge construction, Homestead Act agricultural land reports, past program funds, budgeting, obligating and spending, the purchasing of goods, feed for animals, communications, grazing permits and payments, sheep and cattle conflicts, salaries, trespass reports, fire reports, hunting and fishing regulations, hiring, diaries, search and rescue operations, the six new Washington office districts and their effect, plus several more topics.

Kent wanted reports from each ranger on their current activities and progress.

The day was used and the group ate heartily. Buck and Alva did most of the cooking. No one complained since they knew the consequences.

It was agreed that no drinking would take place the first evening. They needed sharp minds for the next day. The supervisor assured them if they finished what was needed the second day, it would result in a rollicking evening of fun and stories. They did sit outside under the stars and listened to sounds of the night before turning in.

After breakfast, Kent asked for a copy of the property inventory from each district. He said an inspector would be on the forest during August and wanted to check the findings and reconcile them from his list. Kent went down the list and asked for questions. The Office of Property Auditor in Ogden, Utah, recorded only non-expendable property.

The district forester was dispatching half a dozen men to install various timber plots around the forest. These plots could be for a number of research studies and must be protected. For example, one plot could be to test no slash disposal versus piling and burning during a timber sale. Another could be for the rate of growth of certain types of species. They also would be starting a nursery at one of the stations and collecting tree seeds from squirrel caches on the ground and cones from the trees. Some experimental range seeding would also take place and they were looking at tree diseases and insect control. Whoever showed up to do the research needed a place to sleep and transportation, if needed. It was up to each ranger to provide these.

In closing the subject, he stated that if there was an insect attack in commercial timber areas and the trees could not be salvaged, they should be felled, bark-peeled and burned.

Kent gave a short history of the state controlling game animals since 1891. He told the group that each one would be permitted to serve as state game wardens and cooperate with the state, so long as that work didn't interfere with their official duties.

A question arose about the Forest Service reducing wolf and coyote populations by using hunters and poison.

The supervisor responded by acknowledging he was receiving pressure from the local stockmen to reduce these animals' numbers. In the past he had resisted ordering the rangers to place poison in the field, since he was afraid it would effect rodents, birds and even dogs. He also said that on forests where an aggressive program of extermination took place the stockmen were still not satisfied. He noted that even if only one bear, coyote, lynx, wolf, wildcat or cougar were left, it would still be cause for complaint. He couldn't win so he had never started.

Kent mentioned there would be another ranger examination on the forest the following year. He urged each ranger and everyone employed on the district to keep up to date on first aid knowledge. More and more of the public was hiking, riding and driving into the forests. He predicted that search and rescue operations would grow in number and gave the recent example of Buck's encounter with a hiker and his broken leg.

Kent talked about the special Act of Congress that authorized the Forest Service to use national forests receipts to build improvements. The agency came up with a standard cabin. It had two rooms downstairs with the upstairs attic being one or two rooms. It had an outdoor privy. In this log structure the ranger would have to house his wife, family and office. The two married rangers on the Maahcooatche shuddered to think of it. They were content that their stations had separate living quarters obtained several years before the Department of Agriculture took over.

Kent next reviewed the funding for fiscal year 1910. The district forester had agreed with the decision to enlarge the forest with the Long Tom Ranger District and Station, along with the funding for a ranger and guard. They had also agreed to fund the assistant ranger and additional guard on the Lakefield District requested the year before. This meant that Buck would have a full-time assistant and another half year fire guard. The supervisor then asked the group if any of the other district guards or assistant rangers showed a desire to transfer to another district on the forest.

Chapter Five - 1909

Ralph indicated that his assistant ranger, Marshall Hubbard, told him if there was ever an opening on either the Lakefield District or the Canyon Springs District, he would be interested in moving. His soon to be wife Cecilia was from Harmony and wanted to be close to her aging parents. Buck had remembered Marshall and was impressed by the young man's enthusiasm and willingness. It was agreed that Ralph would talk to his assistant and Kent would make the offer.

The supervisor told Buck he had a candidate for a possible guard position but the ranger would be the final judge. The young man would be sent to Lakefield when the meeting was over. Buck said that would be great. He could hardly believe the possibility of doubling the district population.

Funds were also available to build a campground. Alva had located a perfect place. It had been used by many for a hunting camp but was completely unimproved. He had remembered the beauty of Morgan's Meadow and asked the supervisor to request the money to build a campground there. The new campground funds would go to District 5 said Kent. Buck smiled and silently praised his little brother for his achievement.

More talk on funding and budgeting took place. The supervisor warned everyone that the forest had been lucky for a number of years with few large fires. He expected some year that might all change, so he wanted everyone to be on their toes until the end of the season.

Kent also noted that some merchants in Harmony had complained that the Forest Service seemed to favor certain businesses. He reminded everyone to pass around the wealth, if possible. This brought out a response Kent hadn't expected. Tim said that would be fine, except some businesses raised prices when the government was involved, didn't have the items requested, didn't deliver on time and had a poor attitude. Buck didn't say a word. He was very happy that 95% of his requirements were procured in Shadowcreek where there was little competition.

The supervisor brought out a circular issued by the district forester in May, outlining timber policies. The following proposals were made:
1. Greatly increase timber sales program.
2. Have the operators pile and burn all slash on the sale areas.
3. Make planting a high priority job. Definite areas for planting should be located. Selected areas would be planted to conifers and possibly eucalyptus.
4. Broadcast sowing of seed on burned areas should be tried.

There was some discussion on item two as to the effect of this requirement.

Included in the circular was information from the forester's office in Washington.
1. Establish maximum size of Washington office sales to 100 million BF and maximum term as four years.
2. Maximum district forester sales would be 25 million BF.
3. Free use should continue to be liberal. Areas for down material should be established where possible.
4. Experiment planting should be given greater attention. Five year planting plans should be developed.
5. Experiments to determine effects of grazing on reproduction should be undertaken.

The circular also stated that records of Class A and Class B unadvertised sales would be discontinued in the district office. Supervisors would close sales and grant extensions in these particular classes.

Kent went on to say individual forests now had the authority to approve agricultural special uses on areas next to existing homesteads for up to a five year term. Permittees would be required to vacate the land at the end of the permit period if so decided by the forest supervisor. He went on to read from a letter from the district forester setting forth some policy:

> "Applicants may take a special use and demonstrate that area may be successfully used for agriculture, then apply for listing. Time used under special use will be counted time on the requirement for patent. This was mutually agreed on with the Department of Interior. The Forest Service has been criticized for rejecting applications because the Service had decided to withdraw the area for administrative sites after the application had been made. Mr. Pinchot feels strongly that there has been just cause for criticism of the Service for rejecting applications and then withdrawing areas for administrative sites."

Kent asked the group if there were any future administrative sites under study. After receiving a positive response, he told the rangers they must complete them during the summer and post them immediately.

A few more topics were discussed but by then everyone had run out of questions and comments. Kent ended the meeting by saying that the men looked very spiffy in their new uniforms.

The supervisor reminded the group that after supper they would go

Chapter Six - 1910

outside, settle under some trees and have a party. They had completed the brain work the first two days. The next day they would travel to Buck's newly built bridge and campground, where they would eat lunch and leave from there for home.

This was a welcome surprise to everyone. Kent received a three cheer ovation.

After an uncomplaining supper, Buck was asked to pick a good place for a campfire and sleep. From past experience it was assumed several would not be able to navigate back to their previous night's abode. Each one had brought his own drink but that didn't discourage passing around several bottles at a time. Before the stories started, the men traded local tales about recent adventures. With the lubrication process taking place concurrently, in a short time it was agreed the participants were definitely ready for individual performances. Since Buck was the host ranger he was to go first..

Buck stood and told two stories in his usual animated presentation. "As you know, we had a ranger examination on the forest last year. I was one of the people helping out and this is a true story. The first day we spent on written questions and answers and there were supposed to be thirteen candidates. Twelve men showed up at the office, so we walked over to the hall with sealed test papers. Where the thirteenth man was we didn't know but we couldn't wait. Just as the envelopes were being opened, a soldier still in uniform strode in, said he was sorry to be late but he had just walked more that 50 miles to take the exam. I pointed to a seat and said, 'Welcome, you are now our thirteenth applicant.' He looked at me as if I had just shot him. Then he asked if I was absolutely sure he was the thirteenth. I replied in the affirmative. 'I've walked until my feet bled and I'll walk back the same way but I'll be damned if I'll be the thirteenth in anything.' With that he turned, strode out and slammed the door commenting with what we could do with the lousy test."

My second story I'll read. It's a letter I received from a friend of mine who isn't in the Service but thought I'd be interested. Evidently it's a true story, since Ranger Charlie Shaw of the Rocky Mountain National Forest wrote it when he was a dispatcher.

"One morning eight full strings of ten head each were to leave the station for various destinations. The strings were packed and ready to go at about the same time. They were tied up all over the place when Jack Langtree, the station cook asked the packers to come in for a cup of coffee before they pulled out.

Chapter Five - 1909

"While they were in the kitchen tent drinking their coffee, something spooked one string. They broke loose and started bucking, bawling and running through the other seven strings. This caused every string to break apart and stampede. There were mules and horses bucking and running all over the station. Eighty head of horses and mules were involved in this mix up. Some of them became tangled up. Some were down. Most of them stampeded into the woods, trailing packs and equipment as they ran.

"This happened about 9 a.m. Nobody got his outfit together that day. Mules and packs were found the next day as far away as eight miles south of the station. Some were found on Horse Ridge and Twin Flats in the other direction. Some were found still carrying their packs. Others lost their packs, saddles and halters.

"Eventually the mess was straightened out and the packers were on their way. In a class by themselves, they accepted the mishap as part of the work of moving Forest Service supplies over mountain trails."

Buck sat down and there was enthusiastic applause. As always, at one of the ranger story parties, a good swig of the liquid was consumed by all between speakers.

Kent was next. "Two years ago I read a letter from an irate citizen. Tonight I brought with me a copy of a letter I obtained while in Arizona last year. I don't know the supervisor personally but I think his choice of words is extremely interesting. I've been saving this for a perfect time and place and this is it."

He took out a crumpled piece of paper from his pocket and proceeded to read:

Dear Sir:

I had intended to send in my resignation on receipt of my salary check for January, providing that I received no increase, but at the suggestion of Mr. Bronson gave it to him to file with my grievances.

You see, since my salary is less now than it was about ten years ago, after two promotions in the Forest Service, I rather felt that someone was afflicted with the ingrowing salary habit, and it wouldn't be long before my creditors would notice my financial lassitude.

I had received a number of letters approving my work, or at least I took it that way, and I understand my inspectors give me a fairly good recommend, and recommend for promotion, so I do not fully understand just where my promotion caught the locomotor ataxia.

I guess I must have misunderstood, but I thought there was a good possibility of an increase up to $2,800 a year if one could deliver the goods—folks in the Service intimated that I could. I thought I had a bright future before me, but that durned bright future has certainly side-stepped me along the route somewhere, and must be loafing behind.

I was not promoted in 1905, when the transfer was made from the Land Office. I didn't think much about it at the time one way or the other, but when I did get promoted in 1906, I was glad I wasn't promoted in 1905. I was getting $2,371 until my promotion came along in 1906, which gave me $2,200. I know it was a promotion, for my commission from the Secretary of the Agriculture said so right square in the middle of it.

In 1907, I was raised to $2,300; so I am still shy some of the good old salary that I started with away back in September, 1898, with only the San Francisco Mountains National Forest to handle. The fellows on the Black Mesa and Grand Canyon Forest were getting the same amount that I got, but when they fell by the wayside I fell heir to their territory and their troubles, but none of the pesos they were getting. I fully acknowledge your right to assay the intellects of us wood-chucks, and raise, drop or fire; and it is up to me to raise, fall or git, as the case may be. As I didn't get the first (raise), but the second (fall), I thought I had better take the third myself. One can get a heap more money out of a little old band of sheep, or something of that kind, even if his intellect does not average over 30%, with a whole lot less trouble, and retain some friends; but with this job the general public just naturally gets cross if you try to enforce the rules, and if you don't enforce the rules then you get cross; so the Supervisor gets the double cross whatever happens, and has no pension at the end of the game, to sorter ease down his old age when the pace is too fast. While I think a good deal of forestry, I realize that a man can't live in this country and lay up anything, unless he gets a good salary; consequently believe I should go out and make money while I can.

It takes considerable brain fag and wrangling to gather up the

Chapter Five - 1909

$115,581.34 from timber sales, stockmen and settlers, as well as the fag entailed by judiciously expending $23,459.37 in doing the work. I haven't computed, of course, the different amounts connected with the work on the Black Mesa (N.) and Grand Canyon (S.) Forests. I am under the impression that amounts given for the San Francisco Mountains Forest for last year are the largest receipts for any forest, by long odds. I am glad of it, even if it don't count.

I want to thank the different Chiefs for their many kindnesses to me, for I know a feller gets sorter peevish at times when his troubles come in bunches.

I feel mightily relieved at the prospect of seeing some other feller being accused of prejudice, ignorance, partiality, graft, ulterior motives, laziness, salary grabbing, and other such innocent pastimes.

Ten years is a long time to wrangle over the same ground and troubles, then to look ahead to a heap more of it in varieties and quantities, which will assay a heap stronger strain both mentally and financially, and it certainly aggravates one's desire to sorter segregate.

I am glad there will be a bright young man here March 15, to separate me and my troubles, and let me wander away in new fields, where the bleat of the sheep, the height of a stump, the brand of a cow, nor even a special privilege can hop up and fill me with fright or woe. While according to my idea I have not been treated right, I am not carrying away in my bosom any sassy or lacerated feelings, for I haven't time to use them; and, further, I will be in easy reach in case any of the old grazing crews come up at any time.

Very truly yours,
/s/ Fred Breen
Ex - Forest Supervisor

Kent sat down and the place roared with laughter. It was Tim's turn.

Rising to his feet Tim said he had read his story so couldn't vouch for its accuracy but did think it could be possible, knowing what he now knows.

"It seems there was a pioneer ranger, no longer employed, who first worked in 1897 under the old political appointment system. His supervisor ran a saloon and it was his daily custom to go out on the porch of the saloon, look around and then go back and write in his diary, 'Viewed the forest today.' The rangers then had ranches or other occupations and spent little time out in the woods. This particular person spent the first summer in a camp where he thought the forest should be.

"When the boundary was surveyed he found he had meals in the forest and slept outside the forest. Later a third survey was made which showed he slept in the forest but ate outside the forest. A final survey was completed which showed he finally ate and slept inside the forest where he wanted to be, so no other survey was made."

Another swig, another tale. Ralph staggered to his feet and proceeded.

"I have three short stories that some of you may have heard. They may also be apocryphal accounts. It seems there were a couple of outlaws who decided to file a squatter's claim on land having agricultural possibilities and a couple of rangers found out. When the hombres saw the government men they were busy building a cabin and told them to get off their land before they threw them off. The rangers didn't say anything but just sat on the clothes of the outlaws. This went on for several hours as the four men argued and threatened. Finally one of the undesirables told his partner that the bastards won't leave, so they might as well leave themselves. With that, they asked the rangers for their clothes. Underneath them were a couple of 45 colts. The rangers picked them up, emptied the chambers and told the outlaws they didn't want to see their ugly faces anymore.

"The second story relates to a young newly assigned forester who patrolled a vast area and one day came across a group of women who were living on the national forest land and servicing men working on a huge construction project on private land. When he asked them to vacate they told him to 'go to hell.' He then wired the Washington office: 'Undersirable prostitutes occupying Federal land. Please advise.' The reply said, 'Get desirable ones.'

"The third story most of us can relate to. It seems a young ranger was told to go out into the forest and do his job but he wasn't given any instruction as to what that included. For several days he wandered about and got more confused as to what it was all about. Finally, in desperation he sent a telegram and asked what a ranger was supposed to do. After a month of hearing nothing, he received a return message stating, 'A ranger is supposed to range.'"

The group erupted into loud laughter and each man took an extra gulp or two.

Chapter Five - 1909

It was Hugh's time. Bringing out a couple of sheets of paper he said, "I received a copy of a ranger's reply to his supervisor a few days ago and saved it for now. I know the ranger and supervisor involved since they are on a forest to my north. As you know, we all have to write diaries. One day my ranger friend received a letter from his supervisor asking about his lack of detail in his diary.

"He sat down and wrote the following:

Efficiency is a wonderful thing. We all probably try to attain it. Working plans and Schedules of Work have their uses. Diaries come in the Forest Service Scheme. Most field officers in small communities, who try to be neighborly and helpful and at the same time follow their Schedule of Work and keep their diaries up oftentimes have troubles that inspectors don't dream of.

You have no doubt noticed that I have been charging a large portion of my time as Miscellaneous Headquarters Work. I have been bunching the work this way for convenience as that seemed to cover many jobs. To list separately every job of fifteen minutes to a half-hour during a day would make the diary bulky and require considerable time.

During the past season I have never had to worry about finding something to do tomorrow or next week. Instead, I have at numerous times taxed my wits to pick out the important jobs that could be left undone to provide time for doing more important ones. Yet, since you mention it, I can see that a person reading my diary and having no other source of information would most likely get the impression that I was simply killing time, with nothing to do.

It very frequently happens that a day is entirely lost from the plan of work that each of us has. Perhaps I would start in the morning on a job that had been planned in advance for the day and the following is typical of the way it turns out.

As you know, private contractors have been building a road by the ranger station. As I begin work Engineer Smith comes along and requests that I walk up the road with him and inform whether his plan for rebuilding the irrigation ditch which the road builders had destroyed, would be satisfactory. We spend a half-hour looking the ground over and talking over details.

I receive a call to the telephone and spend fifteen minutes getting connected up with my party and five minutes in conversation. I start out to work, impatient at the delay, hang my coat on a post just as a man arrives very much exhausted. His Ford is stuck in the mud on the Fish Creek Hill. He explains that it never acted that way before but his engine is not working right. Will I help him? Sure. I help him out and if we are lucky and do not have to tinker with the car too much, I get back to work and, on looking at my watch, am surprised to find it is 11:45 a.m.

I have just noticed that a bunch of Bar B cattle have broken into the pasture and proceed to saddle a horse and chase them out and get dinner a half hour late. My wife wants to know why I did not split some wood before I went chasing those cattle. I try to explain but get balled up and make a mess of it; then go back to work with family relations more or less strained.

Just as I get my coat hung on the post and my gloves on, Ryan, foreman for the contractors on the road, arrives and would like to borrow my steel tape to measure some culverts. He only wants it for an hour or so. Ed Black rides in on horseback at this time and he feels very badly about the manner in which the Forest Service manages the grazing business. He offers some suggestions as to how we could make things better in his particular case, spends thirty-seven minutes telling me what a bum ranger I am and how the Forest Service is conspiring to put him out of business; gets the load out of his system and goes his way feeling better.

I am called to the telephone to explain to Mrs. White how to corn beef and to Mr. Green what to do for a sick horse. Mrs. White takes up fourteen minutes of my time and Mr. Green exactly eight. While I am thus engaged, Jones' dog chases a bunch of cattle through the fence, tearing down eight panels, and I work until dark cobbling it up again.

I sit down to write up my diary for the day. I begin to enumerate the many things done and decide that if I write all this stuff, that pretty soon I will need help to carry my diary and I am tired and don't feel like writing anyway, so I enter it as follows:

Did miscellaneous headquarters work—unclassified, 8 hours."

Instead of laughter there were cheers as each ranger could relate to the truthfulness of the story.

Alva was last in line for the first round of stories. He stood up and spoke. "This is not a Forest Service story but it could be. A cowboy

Chapter Five - 1909

friend of mine told me about his experience with a couple of mules he was leading on a trail riding his horse. Jake told him that it was important to the mules to establish dominance by being the first one in line behind the horse. Each tried to outrace the other to claim the prized position initially and once there, had to protect his place by fending off the other mule who tried to push and shove his way ahead of him. When the trail became steeper the second mule had to wait for one of the few wide places in the trail before he could attempt his assault once again. At these times the first mule would push his nose as close to the horse's tail as the horse would allow, then furiously stand his ground. In this way he successfully maintained his post all the way to his destination. Jake unloaded the mules and brushed them down before releasing them in the corral. Both animals headed straight for a dust hole and wallowed in it on their backs, then stood up to shake themselves. The humiliated and unforgiving second mule then walked slowly up to his partner, turned quickly and kicked the mule who was first squarely in the ribs with his hind legs, clearly making a statement about his feelings on mule dominance."

Alva was rewarded with much laughter. It was now time for the second round of stories and anecdotes but before starting, an extra few swallows of various brews were consumed. During round two, the speakers didn't stand. Some had difficulty with pronouncing names and places but the rest fell asleep.

It had been two years since the ranger group had a relaxed get-together and the next morning some of the men said they felt that every two years was enough.

After breakfast Kent said they were going to travel west over the rough wagon road that Buck had built to view his new bridge and then go on to his Morgan's Meadow Campground. Hugh and Alva figured they wouldn't return but would continue on to their own districts. The three men who came in the Model T didn't have horses, so they rode in the automobile until they came to the TL&W Railroad tracks. The rails were too high to ride over and the wagon trail got much rougher from there to Shadow Creek. They left the vehicle and walked the two miles. There was much admiration for Buck's ingenuity and workmanship on the bridge. Several said they knew exactly where a like structure on their own district was needed.

They rode and walked to Morgan's Meadow. A combination of the beautiful site and excellently crafted improvements caused the supervisor and other four rangers to pronounce it the best campground on the forest. Luke had arrived earlier, so when he and Buck heard the praise, they felt well rewarded. It was close to noon. The men sat at a table and ate their sack lunch. The two rangers from District 4 and 5 said their farewells and the rest traveled back to Dalton's Station. As the cloud of dust from the Model T faded in the distance, Buck and Luke sat down to discuss their next priorities.

* * *

Something had to be done about trespassing cattle on grazing districts of the Lakefield District. Buck had been too busy to be with the various stockmen during the spring entry. In other years he had made an effort to ride with them as they drove cattle to the grazing grounds. He was able to quell any minor differences right on the spot. With more stock and more stockmen adjacent to the forest boundaries, pressures were mounting to open up the land further. Buck had received several reports regarding stockmen having cattle in the wrong place, having too many cattle or even branding cattle from another owner. He heeded to nip the situation in the bud.

Buck asked for help from his friend, Jug Handley. He was a respected cattleman and one of the few to accept the Forest Service rules and regulations without much complaint. He agreed to help call a meeting at the Dalton station.

Since every stockman had a stake in the outcome, they were all present when Buck called the group to order. He reviewed the past and present situation and said he had read of other forests with similar problems that were solved when the local cattlemen formed stock associations. When one of the men asked how that would help, the ranger gave several advantages.

"All you permittees were furnished prints and written descriptions of your allotments so you could locate your boundaries. Probably the best way to keep your stock on the allotted range would be to hire a range or line rider to move the permitted cattle and salt and keep them on fresh feed. The ranger must be notified of any unauthorized animals in the area. The other thing you could do is build drift fences in a cooperative manner. And finally, it is imperative you do not take matters into your own hands if there are trespassers. It's up to you if you want to form an association or not."

Buck paused and a deluge of questions and comments came from almost everyone. He asked Mr. Handley to take over the meeting. The

Chapter Five - 1909

entire group agreed that something had to be done. Jug was patient and let everyone blow off steam and give his opinion. About an hour later, there wasn't much more to say so he asked if there was a motion to form a stock association. It was moved and seconded with a resounding 90% agreement. Jug Handley was elected as the first president of the Butte Rock Cattle and Horse Association. Within a week, three range riders were hired for the Holcomb Valley, Cuddahy Meadows and Pishi Rock allotments. A major problem had been averted by Buck and Jug.

* * *

The days were getting warmer and the grass and underbrush drier. Lightning had caused several small fires on the Lakefield District. Marshall Hubbard had transferred from the Fulton Ranger District to Lakefield keeping the same title of assistant district ranger. Buck was thrilled when he arrived and promptly assigned him to the Dalton Guard Station. Within a week Marshall had seen much of the district, shadowing his boss as they checked on people, projects and problems.

Marshall was to be married in a couple of months, so Buck figured his mind wouldn't be totally on work. He was pleasantly wrong with his assumptions. The assistant ranger was involved a hundred percent in his dedication to the Service and his determination to do things right. He listened intently and asked intelligent questions. Buck figured he was the lucky one to have such help.

Marshall's horse was a medium sized sorrel named Bucky. This situation caused some laughter for awhile, when he would order his horse to do something and fail to pronounce the "y" distinctly.

The assistant ranger was an interesting man. He had graduated from college with a business degree but preferred a life in the great outdoors, so he left the accounting firm that had hired him, bought Bucky and became a cowboy. At about six feet one, muscular, clean shaven like Buck, he knew ranchers, cowboys, horses and how to survive in the wilds.

Marshall's future wife, Cecilia, was a medical assistant at the Harmony County Hospital. They had met while Marshall was getting his appendix out. She was a cute, baby faced woman with freckles, curly hair and an unbelievable hourglass figure. Although not the jealous type, whenever his fiancée received a flattering whistle, Marshall still couldn't quite get used to it.

Buck, Marshall and Luke became a well oiled three man machine for the district workload. No one complained and they got things done.

The supervisor's promise to send Buck a candidate for his open guard position right after the ranger meeting was a disaster. The young man named Silas Pegler was a relative of a forest supervisor friend of Kent's, who had no knowledge of Silas's past, intelligence or physical attributes. It was a mistake he never again made during his long and distinguished career.

Silas was supposed to meet Buck at the train depot in Shadowcreek. No one got off and a quick check inside failed to find Mr. Pegler. When the ranger reported to the office, Lucy said he was there. Later Buck said that should have sent out some alarm bells. He was also told that the young man had no knowledge of horses but could cook according to his statements. The ranger asked Lucy to put him on the next stage to Shadowcreek and tell him to wait at the stage station.

Since Buck had a couple of hours to kill, he talked to Elroy about renting a horse, then walked with Titus to the post office. It was busy, so the ranger and postmistress talked in spurts as customers came and went. Finally, Buck said he had to go but not before telling Edna that he believed the Lakefield District had a cook. It was too bad he couldn't tell the future.

The stage arrived ten minutes early but there was no sign of Silas. Buck rode around town looking in stores and saloons. Eventually he got to the train depot and there was a young man sitting with a bewildered look. The ranger's assumption was correct; it was Silas Pegler. Silas was quite young, wearing a beat up hat and boots. He had long unkempt hair and was trying to grow a beard and mustache. He was pleasant and smiling, as he shook Buck's hand eagerly.

When they got to Elroy's for the horse, Silas didn't know how to get on or which side to do it on if he could. Patiently Buck showed him how easy it was and they rode off toward Lakefield to rendezvous with Luke and Marshall. It took an extra half hour to arrive. Silas was afraid to trot and when he kicked his horse to go faster it took off in a gallop with the rider yelling and holding on for dear life. This happened three times on the trip.

After introductions were made, Buck asked the newcomer if he could cook. With an affirmative answer, the ranger showed the cook where things were located, including the ice house meat and other frozen goods, utensils, running water and all that would be necessary for the evening meal. Assured that Silas said he could do it, Buck and the other two left the cabin to do some repair work on the barn and corral.

Chapter Five - 1909

The ranger figured he had better check in a half hour and when he did nothing had happened.

"Is anything wrong?" asked Buck.

"No sir, I'm trying to figure out who should sit where at the table and what you want for supper."

"My gosh, Silas, it's been a half hour. As the cook, it's up to you what you want to serve. I've shown you where the food is stored and as for who sits where, we don't care."

"Please sir, give me an idea of what you want to eat."

"OK, get four steaks from the ice house; cook them on that grill over there; cook a vegetable; slice some bread; make some coffee and slice some of that pound cake. You do know how to make coffee? Right?"

"Oh, yes sir."

"I'll be back in a half hour to see how you're doing." With that Buck strode out wondering what he had gotten himself into.

After two more false starts the four men sat down to eat. The ranger reprimanded himself for believing his employee was a cook. That occupation was gone like the wind.

In the days that followed, Silas tried but it didn't matter. The harder he worked at something the more disastrous it turned out. When Buck asked him to feed the animals he threw the food into the same trough as the water figuring they would eat and drink both at the same time.

During one cool evening when the four were sitting outside with a campfire, eating supper, Buck asked Silas to put another log on the fire. He did but forgot to let go and he ended up with soot from one end to the other.

When they went on fire patrol, Buck and Silas separated from Luke and Marshall. The ranger gave the young intern lessons in what to watch for as they rode to the top of Tip Top Mountain. Buck saw the smoke instantly but waited for Silas to see it also. When nothing happened, the ranger pointed toward the rising particles and asked if he had noticed.

"Yes, I saw it but I thought it was a low flying cloud or something," was the reply. They took off in a fast trot, with Silas holding on for dear life. He bounced his way to the fire which wasn't large yet, but could be, if not surrounded with a fire line immediately. Buck instructed the young man to grab a shovel and start moving dirt away from the fire like he was doing. Silas just stood there.

When Buck asked why he wasn't working the answer was, "Well sir, you're doing it right handed and I'm left handed so I can't stand like you are and do it." Buck almost lost his temper.

After a couple of days to make sure the fire was out, they headed back to camp. Early the next morning Buck decided to take Silas to Kent, even though he was sorry for the young man. He had become a distraction, a disruption and a disaster. The ranger's list of Silas' misdeeds had mounted to more than 30. At first it was somewhat funny but now he had become a hazard. Silas had to go and Buck knew it wouldn't be pleasant.

During their ride to town, Buck talked to Silas about the problems. He was direct and to the point in explaining the reason he had to be let go. Silas said he tried to do things right but knew he was clumsy and often couldn't comprehend directions. The ranger gave him some ideas on how he could improve on future jobs. His personality was fine but he needed to mature in his actions and start to think for himself. Mainly, Buck told him he should seek out an occupation more suited to his performance.

It was midafternoon and Kent was in a good mood. The ranger asked the young man to wait in the front office with Lucy. It took about an hour for Buck to go over each item with his boss. At the end, Kent was disappointed but knew his ranger was right. Silas had to go. It was up to the supervisor to do it. After all, he was the one who had done his friend a favor. Buck shook Silas's hand, said he was sorry it hadn't worked out and if he had any questions, Kent would answer them. He had to get back to Lakefield that evening. It was dark when he arrived.

* * *

"Hold still, Buck," exclaimed Edna. "I can't tie your tie if you keep squirming."

"I'm sorry, Eddy. Haven't had one of these things on for years and never did learn to tie it properly."

"There! It's all done. You look so handsome in your suit. The first time I've seen you in one. How do I look in this dress?"

"Beautiful, as always," said Buck. "Cecilia will still be the second prettiest at the ceremony."

"Pshaw, Buck. You know darn well she has the best figure around these parts."

"That's possible, Eddy, but wait until she has one or two children. Anyway, I think we make the best couple, anywhere."

"I'll agree with that, Buck."

It was Cecilia Risby's and Marshall Hubbard's wedding day. Buck was asked to be the best man, so he had to rent a suit and borrow one of

Hiram Lawrence's ties. All the rangers from the Maahcooatche Forest were there.

It was a traditional wedding at the First Presbyterian Church of Harmony and was performed without a hitch as the local newspaper reported the next day. Whether the word "hitch" was used on purpose or by chance, the couple did get married.

It had been arranged for Marshall and Cecilia to have their honeymoon at Dalton Station, where they would also live. Cecilia had moved in some of her furniture and had fixed the house to be a real residence instead of a combination office and bachelor quarters. Buck told Marshall he could take off three days unless there was a fire. After that, he would either come by or phone to plan what they would do the rest of the calendar year.

That autumn Kent asked Buck to help him at the office as before. Without a deputy, he was way behind in work. The ranger tried a little friendly threat by stating he would, if the supervisor would teach him how to drive the Model T. Although Kent knew Buck would help no matter what, he responded by charging him with blackmail. This dueling went on for some time before Kent finally agreed to teach Buck the finer points of driving, just as he had agreed to months before. Lucy couldn't help but overhear the conversation and could hardly refrain from laughing. She loved it when the two men she admired most had one of their bantering boyish behavior exchanges.

* * *

The days were drawing shorter and the hardwood leaves were turning into multiple tints and shades of colors. Luke was off the payroll and Buck had closed Lakefield, since he would be in the Harmony office during winter. He still planned to patrol his district with Titus but now that Marshall was on board he felt a little easier about being out of touch.

The year ended with Buck and Edna together at both Christmas and New Years. 1910 was to prove a jolt to the Forest Service, regarding fire management presuppression and suppression. New ideas, equipment and innovations were coming.

1910

Forest Fires

There's a roarin' fire a ragin' through our splendid timbered slopes,
    And we're fightin' it like devils—withnorhin' but some hopes.
Just a smoky sky above us and the cinders 'neath our feet
    And no peltin' raindrops fallin' to make our souls more sweet;
With no bed of downy "suggins" waitin' for our needed rest,
    With no chuck at all to feed us—and this surely ain't no jest!
And the pay is a hundred dollars—Oh, the rangers' living swell,
    And we like this forest business, but—Fires is hell!

J.D.G.

CHAPTER SIX

1910

On January 7th, Forester Gifford Pinchot left his home to have dinner with Philip P. Wells, a Yale classmate and close friend.

Thornton T. Munger was present and wrote in his diary: "He greeted the company very graciously and cordially took a seat by Eleanor Wells on the lounge. He had two envelopes, one opened, the other not, which had been given him by a messenger on leaving his house. Soon after seating himself he said very cheerfully—'This tells me I've been bounced. Excuse me if I try to find the reason.'

"After reading to himself the two page letter, he said—'possibly you may care to hear what it says,' and proceeded to read the letter from President Taft. After reading his letter he opened that from the Secretary of Agriculture dismissing him and read it aloud at the first reading."

The next morning at the old Atlantic Building, Mr. Pinchot assembled all the Forest Service people and bid them goodbye, saying, "Stick to the Forest Service."

On February 1st, Henry Solon Graves, Dean of Forestry at Yale, replaced Pinchot as forester of the Forest Service.

The field personnel were stunned to hear the news. Pinchot had been their one and only leader since 1905 and before. Changes would be inevitable but they continued to work as always, in the same hard driving manner. Kent and Buck discussed the ramifications but had heard that Graves, although lacking a sense of humor, was a good man.

* * *

The district forester sent Kent and other supervisors a letter establishing the policy of completely marking timber sales up to 160 acres before advertising, so purchasers could see what timber was available. On advertising of more than 160 acres, at least a quarter section was to be marked in advance. The marking could be temporary and provisional, by typing a string around each tree instead of a breast high blaze and brand below the cut lines. This way, if a prospective purchaser withdrew, no permanent marks would be left.

The letter went on to say, "Remember that an area improperly marked will result in a condition that cannot be remedied. It will stand as a permanent monument to inefficiency or bad judgment. On the other hand, an area properly marked will serve as a lasting example of the marker's skill as a forester."

A second letter from the district forester received an immediate reaction from Buck. It was a list of furnishings that could be purchased from government funds for use at ranger stations. He wondered aloud where in heck had this policy been for the past few years when he paid for them out of his own pocket. The list consisted of:

1 cook stove
1 heating stove
1 office table
Chairs for office table
Canvas cots for use by transient forest officers
Window shades
Screen door and windows
Suitable filing equipment
1 typewriter
1 typewriter stand
1 bookcase

The Forest Homestead Act of June 11, 1906, had created a huge workload for field personnel. Buck and his fellow rangers had surveyed, submitted and processed dozens of claims under the act. Another act of Congress titled the Forest Allotment Act of 1910, creating Indian allotments, also caused a vast amount of work. The first act specifically stated it was for white men only. There had been many Indian families occupying the lands prior to the establishment of the Forest Reserves. It had taken four years but Congress, under pressure, now allowed Indians to also claim land up to 160 acres, provided they had a minimum of 5/16th Indian blood and could prove they had occupied those lands. Depending on the type of use, the number of acres consisted of a graduated scale from 40 to 60 and so on.

Indians were given serial numbers through a registry at the Bureau of Indian Affairs. If they did obtain an allotment, it could be passed on to their children. The BIA received a trust patent of 25 years for the land. If an individual occupied the land under trust patent for 25 years,

Chapter Six - 1910

it became a fee patent and was owned outright by that person. These claims took decades to determine. If an Indian was a tribal member of a reservation, he was excluded from obtaining a separate allotment.

* * *

It was springtime. The wet winter had caused the grass to grow fast and dense. The railroad had not cleaned up their right of way, even though the Forest Service supervisors had written several letters indicating their need to comply.

The telephone rang. Lucy answered and since Kent wasn't in, she asked Buck to take the call. Something in Tim Westgate's voice stopped the urge to start with a sociable chat. Tim almost shouted that a slow moving freight train had gone through his district and he had received reports of it throwing off hot embers and starting several fires. He told Buck it was coming his way. Buck, in turn, asked if he needed to dispatch anyone to the fires. Tim said he could handle it from his end but Buck better be prepared to fight his own fires along the railroad's right of way. The ranger ran to the Harmony depot and calmly told the railroad dispatcher, Tucker Foss, what was happening. He said it was urgent to stop the train before it started more fires.

The dispatcher said, "Sorry, but I don't have the authority. Only the top brass can stop a train on this line."

Buck didn't wait to hear more. He ran back to the office, started Kent's Model T, drove back to the depot, parked squarely on the main track and darted into the railroad office, this time yelling there was an official government automobile on the main line and they had better stop the train NOW! Tucker almost had a heart attack. He frantically telegraphed a message to headquarters with a quick rundown of what was happening. Hearing the train whistle, he grabbed a red flag and bolted out of the office and down the track to the south. The locomotive stopped just short of the depot and the Model T. Buck could see the hot embers spewing from the smokestake and ran to tell the engineer the reason he was stopped.

The engineer, Tucker and Buck walked to the depot office. A reply to the telegram said to let the train continue, fires or not. This made Buck furious. He threatened to get the local paper involved and to sit in the vehicle and let the train engine hit him as it moved forward. He also said the railroad already would be billed for the suppression costs to the south and that to continue might cause a major conflagration that could cost additional thousands of dollars.

In a panic, Tucker then called headquarters on the telephone. He wanted to talk to someone in command. He needed a name and direct order to protect himself. Tucker listened for a few moments, then handed the telephone to Buck.

"Hello, this is Buck Stonewall at the Maahcooatche National Forest."

"This is Collier Travers, General Manager of the TL& W Railroad. I understand there is a problem at Harmony."

"There sure is, Mr. Travers. Your engine is blowing red hot carbon particles from its stack and has already caused several fires to the south on national forest land. I don't believe you want it to continue north and cause dozens of more fires, which would add to your suppression costs in a huge way."

"I understand you have placed an automobile on the tracks and won't move it."

"That's right and I don't intend to move it until I have your word that you'll either fix the engine or replace it. You should understand, Mr. Travers, that your railroad will take a black eye in the press if you continue to knowingly cause wild fires, and as a minimum, I'm sure there would be Congressional investigation." Although Buck was bluffing on this last statement he threw it in for emphasis.

"Well, Mr. Stonewall, your statements make sense, even if I don't like your methods."

"Sorry to have to do it, Mr. Travers, but your policies caused it. Do I have your assurance that the current engine stays put until it's fixed or replaced?"

"All right, I agree. Let me speak to the dispatcher."

Buck handed Tucker the telephone and told the engineer what had happened, while the other two were talking.

"Headquarters is sending a replacement engine immediately. The present engine will be held at Harmony for repairs," said Tucker as he hung up.

The engineer pulled the train to a siding, unhitched the engine and moved it to a service area. Buck drove the Model T back to the office. Kent had returned and knew nothing of the episode. As Buck told the story the supervisor changed facial expressions several times. At the end he didn't say a word for a few moments.

Finally he spoke, "Buck, I've never heard of a story that comes anywhere near what you just did. You have the damnedest way of getting into and out of problems than any other person I've ever known. I

Chapter Six - 1910

have to congratulate you for what you did. But I do have one question. What would you have done if the train hit my car?"

"That's a purely theoretical question, Kent, and I refuse to answer it 'cause I don't know."

"You always know what you are doing Buck, so I won't belabor it."

News of what the ranger had done spread quickly. Calls came into the office supporting the action. Kent received a phone call from the district forester and the local paper did a front page story. All the attention flustered Buck, who asked Lucy to direct all calls and visits to Kent for a few days. He would be busy talking to Luke, now back on the payroll, and Marshall, regarding fire plans for the coming season. He told them, from all indications, it looked like a rough one.

* * *

One day Kent came into Buck's office with a big grin. "Buck, I've got some great news. You'll be able to return to Lakefield in a couple of days after our new deputy forest supervisor arrives. His name is Stuart Brewer and I'm told he is a good man, although new to the Forest Service."

"That's great, Kent. If you want me to overlap a few days to get him acquainted with things, I'd be happy to."

"Hoped you'd say that, my friend."

The day came. Kent met Stuart at the depot in his Model T and returned to the office. Introductions were made and the three men sat in the supervisor's office for a couple of hours.

Stuart Brewer was a recent graduate of Yale University. He was considered a top student with unlimited possibilities. He had requested a western forest and wanted to learn from the ground up. He talked slowly and deliberately. Of average height and weight, he had curly brown hair and a thin face. He said he was willing to get his hands dirty and his feet wet. He needed a horse but was no horseman and asked if one of them would help him buy one. Buck readily agreed. He liked his new boss because of his attitude and willingness. For the next three days, the ranger and new deputy worked together in the office. Buck found a horse named Buddy and helped Stuart purchase his saddle and field gear.

It was time to return to Lakefield District. Buck saw Edna before he left with Titus and Stub loaded with food and supplies. He rented four horses from Elroy for a load of lumber. His plan for a fire watch was to become a reality.

Buck, Marshall and Luke met at the station on the agreed date. They spent a day repairing, cleaning and putting the station back into working condition. The phone line was already fixed as Buck had requested.

With all the necessary tools, lumber and supplies, they rode west toward Florin Peak. Scouring tall timber, they found a suitable Douglas fir for a lookout platform viewpoint. Next they felled two fir poles and dragged them to the site. After cutting the lumber to proper lengths, they proceeded to make a 50 foot tapered ladder, with the poles as the rails. The poles were five feet apart at the ground and two feet at the top. Buck donned climbing gear and told the others to lean the ladder against the tree and push as he pulled from the top of the rope. With the lower end leaning away from the trunk, Buck nailed and tied the upper end. They built a second ladder with no taper. Although it was only 20 feet, it was heavy. Buck and Luke climbed first and both pulled as Marshall pushed from the first ladder. They fastened it directly to the fir tree. It took several hours to build a small, secure platform around the tree at the top of the second ladder. It was now possible to view the complete panorama above most of the other trees. Using binoculars, they could see locations of smokes with comparative ease. Before the season ended, the men built one more tree lookout and planned a third for the following year.

Buck had kept track of his grazing permittees and succeeded in checking for arrival dates, numbers and proper locations. The Association seemed to be working well for all parties. He hadn't been hit with Indian allotment problems yet but the law had just passed and he expected activity soon.

* * *

One gorgeous summer day Buck and Titus were alone, checking out some trails and signs, when the ranger heard a clanking sound to the north. He stopped and listened. It continued off and on for several seconds, so he left the trail. In a couple of minutes, he watched two men digging feverishly as their shovels hit rocks. They jumped as Buck approached quietly and asked what they were doing.

"Digging for artifacts," came the reply. "Arrowheads and trade beads are quite valuable on the market."

"Have you found any here?"

"Yes, we figured this to be an old burial site and we've hit pay dirt."

"Are you archaeologists?"

"Nope. Just amateurs pursuing a hobby."

Chapter Six - 1910

Buck knew the law. The American Antiquities Act of June 8, 1906, provided protection for objects of antiquity, plus it authorized presidential proclamations to create national monuments. The digging for arrowheads and trade beads was not legal on national monument lands but was legal on national forest lands. He thought to have some fun, anyway.

"Do you know that there are some states which don't allow digging up Indian burial sites?"

The men stopped digging. Buck had their attention.

"Gosh no! Does the state we're in now have such a law?" asked one.

"I'm not with the state. I'm a federal employee in the Forest Service," replied Buck.

The two men looked bewildered. They discussed between themselves their next move. Buck bid them adieu, with the admonishment they better check things out before continuing. He thought to himself what a ridiculous situation, to allow indiscriminate digging of artifacts in the national forests.

* * *

Buck and Edna made plans for a picnic and campout at Morgan's Meadow Campground over the July 4th weekend. Edna hadn't seen the site since Buck and Luke had made the improvements. She bought and packed the food, along with a small tent in case of rain. Titus, Stub, a rented horse, Molly, Buck and Edna headed north via the Shadowcreek road.

Marshall was on duty that weekend, so when they stopped at Dalton Guard Station, only Cecilia greeted them. She was very happy to have company. She said she hadn't realized it would be so lonely. Edna comforted her and gave some pointers on how to spend time creating, gardening and educating herself.

Luke and Marshall had been working on the wagon road from the main road to the bridge. It was now fit for automobiles.

When Edna first saw the bridge, she couldn't believe it. She marveled when Buck told her the blow-by-blow details of its construction. It was apparent it had been used by many people and animals during the past year. There were a few other families at the campground. Buck and Edna found a secluded table and bench and placed Stub and Edna's horse in a nearby corral with food and water. Titus was left on his good behavior, as always. Molly had run most of the way and lay down exhausted.

Edna had purchased several items that Buck had never seen before.

When she opened a tin full of bacon, he was amazed. The ranger always drank water from the nearest stream or spring. The carbonated beverages in glass bottles from England were something new. They had rounded bottoms, so the drinker couldn't set them down. Each bottle had to be emptied or propped up somehow. With a flick of the thumb, the cork wired down on top of each bottle could be removed. She had corned beef in tapered cans with a key wind strip. Instead of digging out the contents, it would easily slip out for slicing. For fruit they had canned peaches. At that time, the U.S. had what was called sanitary tin cans made of 100% tin. The outside overlap was soldered with a rubber gasket seal in the seam which prevented lead poisoning. They opened the tin cans with a knife.

Edna and Buck ate slowly, talked and enjoyed their meal. Cookies were dessert. Edna saved her baked gems[22] for the morning meal. Campfires flickered here and there. The spaces were far enough apart to enjoy privacy. Eventually Buck brought out his harmonica and Edna sang softly. The evening brought them closeness and contentment.

* * *

That summer and fall, there were more fires on the Maahcooatche than in any of the five years before. Rangers went from one district to another helping out. Several dozen men were hired during the season to fight fires on various fronts. Buck, Marshall and Luke had little time for anything else after August.

One morning they were completing a separate fire cache building at Crescent Ranch when Sarah received a phone call. The Barnesville District had a large fire to the west near the Crescent River. Some men were dispatched from the Long Tom side and others were coming in from Barnesville to the north. The three Lakefield men were asked to head west along the river and do mop up work on the east and south sides. Leaving the horses behind, they took tools and some food for four days.

The terrain was steep down into the river from both sides and the forest was thick with trees and undergrowth. The fire had burned to the north side of the water, about five miles beyond Crescent Ranch. One of the many creek tributaries had stopped it from moving further east. Trudging through warm ashes, they didn't stop for another two miles until breaking out into an untouched meadow. It was apparent the fire was heading northwest uphill, so they formed a line along the burned area to quench any hot spots. This went on for about an hour, and they, working 100 yards apart, were making good progress. There was a half

mile of untouched trees between where they were working and the river. However, there was still a great deal of unburned ground vegetation within the burned over area.

Marshall noticed it first. A little change in the wind direction with some swirling dust. He stopped and yelled at Luke down the line, who in turn yelled at Buck. They all stopped to reconnoiter. Some more swirling dust and a breeze really got their attention. Then it happened. The wind changed 180 degrees almost instantly and they could see the fire in the distance pick up flame and head their way burning the unspent fuel. They all knew what that meant. Fire can and does burn downhill, and with a good wind, can move faster than a running man. If it got to the unburned timber area they were cooked, unless they could outrun it to the river. It would be close.

All three men dropped their tools and took off. Buck was nearest to the river. With his long strides and quick turns he seemed to glide like a frightened deer dodging trees and forest floor debris. Luke was next and Marshall third. Halfway to the water there was a sudden roar. The flames had torched the tall trees and were gaining on the runners.

As he was moving, Buck was thinking as he always did in emergencies. The only safe place was the river but did he have time to remove his boots when he hit the bank. If not, they would surely fill with water and weigh him down, making it almost impossible to reach shore 50 feet to the other side, even if he was a powerful swimmer.

He could see the raging water from 40 yards away. Quickly looking for an open spot, he spied a small sandbar in the middle about 10 feet downstream. Luckily his boots had enough high hooks at the top so he could rapidly crisscross the laces, loosen and remove them from his feet in a few seconds. Taking a chance he swung each boot toward the sandbar, a distance of 25 feet, to the center of the river. It was a six foot by 15 foot target and he managed to bounce them on the near side before they came to rest on the far side, inches from the water. By that time, the fire had arrived and the other two men were moving on adrenaline. Buck dove in above the sandbar. By diving, he cut the distance in half and with four or five strong strokes made it to the island. Next he tore his shirt off with buttons flying. Jerking his pants off, he saw Luke jump in with everything on, so he aimed where he thought his fire guard would be carried and dove in. Luke was a good swimmer but not good enough to stay afloat long with heavy saturated clothing and weighted water filled boots. Buck guessed right and grabbed Luke's arm, telling

Chapter Six - 1910

him not to resist but to try and kick and let him do the arm strokes. At the last second, the two men hit the lower end of the sandbar. Luke lay exhausted, half in and half out of the water.

Now it was time to get Marshall. Buck had seen his assistant jump in as he was helping Luke. There he was, floating swiftly past the little island, just out of reach. Marshall was flailing his arms and yelling. Buck didn't hesitate and again dove into the water. With a strong stroke and vicious kick, he finally caught up with the now spent Marshall. As he had told Luke, Buck hollered at his assistant to cooperate and hold on as he swam toward the south shore. It worked. The ranger grabbed a branch sticking out about three feet. Marshall crawled over his body, grabbed the same branch and pulled himself to shore.

All three men were near empty. They had run a half mile, jumped into cold, fast moving water with their clothes on and come out alive. Without moving or talking, they lay still for a good ten minutes.

Buck was the first to rise. He was naked except for socks and underwear. The current had taken the two men down river so far they could barely see Luke, now sitting in the middle of the sandbar. Moving slowly upstream to a point across from the guard, Buck hollered for Luke to take off his boots and clothing. He was to tie the boot strings together around his neck, with pants in one boot and shirt and socks in the other. This would help prevent the boots from filling with water. Next he was to jump in and swim to a long branch Buck and Marshall would hold out for him to grab downstream. He was to wait until Buck's signal. The two men found a good stout, easily grasped smooth branch about 12 feet long. They held the shore end and motioned for Luke to come ahead. It worked as planned.

Now it was Buck's turn. His boots and clothes were on the sandbar. They were miles from anyone else. It would have been almost impossible for Buck to walk without a foot covering in such rugged country. Telling the other two to stay where they were, he headed along the shore to a point about 150 feet to the west of the island. He dove in and swam easily to his belongings. Repeating what he had told Luke, the ranger made shore successfully.

The fire on the north side had burned to the river. Winds had calmed down but there was concern that the river fireline might be jumped. It was getting late; the men were exhausted, cold, wet, bruised and bleeding but glad to be alive. Their hats were gone. They had no food or tools; their clothing was torn and it was foolish to travel in the dark.

With the fire threat still real, they found a level spot southwest of the sandbar. During the night an eerie glow shown at places to the north. Sometimes crashing trees were heard and splashing water, as if an animal was trying to escape.

After a fitful night of little sleep, the three men double-checked for fire on their side of the river and finding none discussed what to do. Although they were on the Fulton District, there were no trails to Fulton Station in the vicinity. There were also no trails along the river. Figuring that Crescent Ranch was about the same distance as Fulton, they slowly staggered, struggled and stumbled their way east, remembering the nearest bridge was at Southcott's, two miles below Luke's place.

It was mid-afternoon when Sarah Parley looked up from taking in the dried clothes and saw three bedraggled and filthy men slowly moving along the opposite bank. When Luke yelled, she could hardly believe it was him. Finally, she understood. Matthew and Mark were to ride Baron and Bucky, with Titus in tow, without saddles to the Southcotts. They would meet at the bridge.

Sarah asked them to leave their boots and clothing outside. If anything was worth saving she would wash it in the morning. In the meantime the three men could shower outside and try to fit into some of Luke's clothes. It was awhile before any of them could be persuaded to talk about their harrowing adventure.

The men's pants could be saved but the shirts and socks were trashed. Luke was smaller than the other two, so when they tried on his shirts, they looked like giant elves with short sleeves and midriffs exposed. It was a hilarious sight and they all laughed, an emotion that was sorely needed. Buck told Luke to take a couple of days off unless there was a fire call. He and Marshall rode hatless on Titus and Bucky toward Shadowcreek.

It was Saturday so Buck also told Marshall to take a couple of days off. The assistant had clothes waiting at Dalton Station. Buck needed a complete new outfit but stopped at Edna's first. She answered the door and gaped.

"What happened, Buck?" she asked, barely able to hold back a laugh.

"I know, I know, Eddy. I'm a mess from the waist down and look funny from the waist up. It's a long story and someday I'll tell you but what I need now is a place to sit down and some new clothes."

"Of course, Buck. Come on in and rest. I'll make some hot tea and we can go down to the mercantile before it closes. I better tend to those cuts and scratches."

Chapter Six - 1910

Her place felt like a second home to Buck, so he took off his shirt and sat while Edna heated some water and rubbed his neck and shoulders and applied some first aid. She continued with the conversation.

"Buck, I want to ask you a theoretical question. If you were married and had a son, what would you name him?"

The ranger looked at her curiously but answered, "That's easy Eddy. He would be Thomas after Thomas Stonewall Jackson."

"I thought so."

"Now let me ask you the same theoretical question. If you were married and had a daughter, what name would you give her?"

"I'd want her to be named after my mother, Mary."

"That's a good solid name, Eddy, but remember we only said theoretical."

"Of course. I'll get the tea and then if you want, we can buy you some clothes."

Buck was torn between riding Titus to the store without a shirt or wearing Luke's. He decided on the latter, even if it caused laughter and questions. Edna had wanted to help Buck buy some clothing for some months. It was a perfect time, she thought, for him to stock up on several pairs of pants, shirts, underwear, socks, belts, suspenders, hats and kerchiefs. He wouldn't buy more than two of anything and said absolutely no when she suggested a new pair of boots. The pair he had on would dry and they already fit. New shoes of any type were a pain to break in. She knew when to accept his decision, so made no more of it.

Buck didn't feel he could take any time off. He needed to write up the last fire report and several others that had been put aside. He also had brought with him from Crescent Ranch other paperwork—reports, diaries and analysis that were needed at the supervisor's office. Buck and Edna had supper that evening at her house. After helping with the dishes, he excused himself and aimed for Harmony and a comfortable bed at the Harmony Boarding House ten miles away.

As he rode, his thoughts turned to Edna and their conversation about naming children. Even though she had said it was theoretical, why had she brought up such a strange question? He also had to admit truthfully to himself that he enjoyed her helping pick out his clothes. It was the first time they had actually shopped together for more than just food. They were troubling thoughts, as he sat relaxed on Titus.

Buck worked all day Sunday in the office, had a meal at the Golden Restaurant and a beer in the saloon.

The next morning, as usual, Lucy and Buck were the first two at the office. Stuart and Kent arrived almost at the same time after 8:30. They were surprised but pleased to see Buck and asked him to bring them up to date. Leaving out the river experience, he answered their questions and submitted his reports and analysis.

"Thought you might be interested in a letter that Garth sent me, Buck," said Kent.

"I sure would like to see it."

"Here it is and it's almost impossible to believe."

Buck had been out of touch with the news in the West. Big fires were raging in Montana, Idaho, Washington, Oregon and Northern California. The local paper didn't go into much detail, so Kent was glad when he received the letter from Garth the previous Friday. Buck opened the letter. There was a cover letter.

Dear Kent,

As you know my forest is burning. We have a catastrophe. I thought you might be interested in an essay the Forest Service's T. Shoemaker wrote about the fires. It's a long one so I'm enclosing a shorter version. This is a very challenging forest to be running. I miss you guys and your high morale and close friendships. Say hello to everyone, especially to Lucy and Buck.

Best regards,
/s/ Garth Kimball

Buck turned to the next sheet.

"This awful holocaust of August 1910 snuffed out the lives of more than 80 firefighters and laid waste half a million acres of timber. This was not just a single fire at its beginning, but a sudden breaking away of many fires that had been burning for days. Men were on these fires or cutting their way to them when a tornado-like force struck and sent them roaring and spotting ahead, fanning sparks into blazes and blazes into crown fires that joined other fires to form an almost solid front as it crossed the Bitterroot Mountains into Montana. It consisted of overheated air that swept up from the desert-like plains of central Washington and was almost entirely lacking in humidity.

"A few lines about the men who, from day to day, came straggling in out of the blackened waste, weak and emaciated from lack of food, feet burned and skin blistered, clothing in shreds,

Chapter Six - 1910

and faces bewhiskered and begrimed to the point of making them unrecognizable. The strong helped the weak up the steep slopes, over the down logs and through the roughest spots, but even they were scratched, bruised and limping at the finish.

"Bodies were found widely separated—one here, two or three there, several close together elsewhere. Mostly they were along trails which they vainly hoped would lead them to safety, or in the beds of streams where they had submerged their bodies as the only chance of survival, only to be suffocated or scalded in the sizzling water as the burning embers dropped in around them.

"Heroically, and methodically, the search went on until all were accounted for. But not all could be identified and some were not claimed by anyone, since they were transients with no next of kin known. It is gratifying to know that a sightly plot of ground was set aside for the burial of all the men whose relatives preferred it, as well as the unidentified, in the cemetery at St. Maries, Idaho. It was appropriately monumented and is scrupulously tended as a mark of respect to the men who, in life, essayed to save the forest from destruction by fire but were themselves destroyed by it.

"Among the bodies recovered were those of 5 men taken from the shallow tunnel into which the heroic Ranger Pulaski took his crew of 40 men as the only chance of survival. The tunnel's entrance was at the bottom of a canyon whose slopes on either side were very steep and heavily wooded. As the fire passed over, great trees, uprooted or broken off by the gale, tumbled or slid down, creating a veritable furnace around the mouth of the tunnel that exhausted all the oxygen. As breathing became difficult, the men instinctively fought to get out. That would have meant certain death, but Pulaski held them back at gunpoint and commanded them to lie down and suck air from the damp floor of the tunnel. Finally, quiet reigned and Pulaski lay down in the most exposed position, the last he remembered until several hours later when the fire had pretty well burned itself out. He was awakened by men crawling out over his body and heard one of them say, 'Too damned bad, the ranger is dead.'

"As might be guessed, all the men had become unconscious. All regained consciousness but five, and after all efforts to revive them failed and it became light enough to see to find his way,

Pulaski led and helped the others down over or under the charred timbers and around the boulders that had tumbled down to obstruct the trail. At last they reached Wallace, Idaho and to its citizens who knew their approximate whereabouts and had given up all hope of their escape, it was like seeing them rise from the dead."

As Buck finished reading he couldn't help but reflect on his recent ordeal with Luke and Marshall. He let out a big breath and felt lucky to be alive.

A few days later Kent received a letter from the district forester.

"Under a recent decision rendered by the Solicitor of the Department, expenditures incurred for medical aid for employees of the Forest Service who were injured in fighting fires and for the disposal of bodies of our employees who were killed while fighting fire cannot be paid from official funds. We have felt that those of our rangers and temporary employees who were injured in fighting fire should not be required to meet by themselves the cost of medical attention and hospital services for injuries received on official duty. We also felt that some proper and fitting disposal should be made of the bodies of the 74 temporary employees who lost their lives while engaged in fire fighting. Wherever such bodies were desired by friends or relatives, we have arranged for their shipment and have assumed responsibility for expenses incurred for this purpose. It is our desire that the unidentified and unclaimed bodies be buried in simple plots at our ranger stations, which can be permanently cared for and protected and which the circumstances under which the lives were lost can be commemorated in a fitting manner. We have felt that the Service as a whole would be glad to meet the cost of caring for the injured men and disposing of the dead in accordance with the arrangements outlined.

"The American Red Cross Society has contributed $1,000 toward the cost of medical attention but this fund is wholly insufficient even to meet expenditures for that character alone. At least $3,000 in addition will be needed to care for the injured and disposition of the dead. The members of the Service in this District are joining in raising a fund by subscriptions not exceeding $1 each. Please present this matter to your staff."

Chapter Six - 1910

Supervisor Bolton immediately wrote a letter to the five districts.

"The following letter just received from the district forester is self explanatory and I trust it will meet with your approval. If all of you who desire to assist in this matter will forward $1 to this office at once, it will be forwarded to Mr. White, District Fiscal Agent. Thank you."

* * *

Buck, Luke and Marshall walked and rode to several more fires during the fall months. Sometimes they were alone, sometimes with each other and once they had a crew of a dozen locals helping. In all situations they studied the terrain, fuel, moisture, temperature and weather patterns intently. They had learned the unpredictability of fire behavior.

One fire came close to Harmony. It started with a lightning strike on the Canyon Springs Ranger District five miles from town. The winds carried the fire front north and east and every able bodied person in town was recruited to help build a fireline along the south side. Buck and Tim and the men from their districts were there. So were Kent and Stuart. Kent asked Buck to be fire boss for the forest area. The captain and men of the town's volunteer fire department helped build fireline but stayed with the town to protect the structures. Buck had about 20 men under his command and after receiving as much information as possible from scouts and line personnel, he decided to take a chance and initiate a backfire from the fireline.

The calmest winds were at night, so Buck ordered a line of men to ignite fires in the dark. He knew if it failed, the town could go up in flames. He also knew if he didn't do anything, the town could still go up in flames. The initial backfire saw flames leap into the air and several of the townspeople ran to gather their belongings. Most of the population had never seen a wildfire up close. It was much scarier than the structure fires they were used to. Spot fires landing across the broad fireline were promptly extinguished.

Gradually the threat to the town lessened and by morning the vegetation had been burned between the main fire and the backfire. Mop up continued for several days and nights. Buck, Tim and the rest of the two district personnel catnapped for an hour or two each night, ate sparingly and didn't think about washing.

Afterwards Kent graciously invited each Forest Service employee to wash and cleanup at his home in town. He admitted having a cleaning lady come in later.

Eventually fire season ended. Luke went to his ranch while Buck and Marshall prepared for winter work.

During the five years Buck had been in the Forest Service he killed more than fifty rattlesnakes. He never went out of his way to kill these creatures but only if he thought it necessary for general safety or immediate expediency. Snakes kept the mouse population down around the station but he knew his own animals came first. He learned rattlesnakes held their rattles high above water as they swam creeks and lakes. The biggest snake he killed was 44 inches long, had few rattles and was quite old. During those five years he had never heard of anyone or any horse or mule being bitten by a rattler on the Lakefield District.

One of Buck's favorite campfire stories, which he told when youngsters were present, was the time two friends of his were walking up a trail not looking down but viewing the magnificent scenery. The first friend walked right over a rattlesnake. The second friend's foot skidded on an unfamiliar surface. At this he went straight up for several feet, then started swinging his knife and had the snake chopped into sections before his feet returned to the ground. That evening they cooked the sausage shaped parts and said it tasted like chicken. This would be followed by several more animal tales that were always entertaining but stretched the imagination, even though the audience loved them.

Buck hadn't had any close encounters with cougars since Edna shot the one charging at the station several years earlier. He did come to recognize some of their habits and movements. They seemed to prefer deer meat, although they did occasionally take smaller prey such as ground squirrel and porcupine. They were adept at turning a porcupine on its back to expose the soft underbelly. Cougars also preferred making their own kill instead of taking the role as a scavenger. Buck noticed they almost always covered the partially eaten carcass with leaves, twigs and other debris to hide the remains from other animals.

One morning rounding a bend in the trail, Buck met a large female cougar accompanied by a pair of half-grown kittens. Buck and the cougar eyed each other for a moment before the female turned and bounded into the forest. The kittens remained as their mother disappeared and were soon back to scampering about. He learned that seldom do cougars attempt to defend their young from man to the extent the female bear would.

On another occasion Buck saw a cougar about 50 yards away, stalking some animal. Since he was downwind he watched for several min-

Chapter Six - 1910

utes as it lay still on the ground. Leaving the trail on foot to do some sign checking, he returned more than an hour later to find the animal in the same position. He knew that the cougar's inadequate lung capacity didn't allow for a long chase.

One of the most spectacular wildlife scenes Buck ever witnessed was when Titus and he were traveling along the bottom of a large timbered canyon split by a fast running stream. A huge six-point buck[23] was drinking, so he stopped to view the fine specimen. Then from the corner of his eye he saw a cougar make his way along the other side of the water and crouch as soon as the buck was detected. Buck and Titus stayed motionless for some time as the deer slowly made its way to within striking distance of the big cat. A running jump placed it on the buck's back and the battle was on. The fight was awesome. Debris, hair and blood was thrown around wildly. The cougar was the eventual winner but lost some hide and gained some bruises.

* * *

From the time Peter Blodgett told Buck he planned to sell his ranch within the next four or five years, the ranger had stored that thought in the back of his mind. It was four years ago and to his knowledge, Mr. Blodgett had not put it up for sale. Buck needed to see Luke at Crescent Ranch about fire tools and an inventory, so after bidding the Parleys goodbye, he rode down the road toward Peter's place. The old man answered the door and was extremely happy to see the ranger. Following the normal pleasantries, Buck asked Peter if he still planned to sell.

"Absolutely," was the reply. "As soon as I can. I'm not able to take care of much any more."

"I'm sorry to hear that, Peter," answered Buck. "I'd be interested in buying it, if the price is right."

"It's all yours for $1,500. I figure that's as much as I'll need to live out my life with my widowed sister in Harmony."

"That's a generous price all right and it's within my budget. Shall we shake on it right now?"

"You bet. I'd hoped it would be someone that I knew would take care of the place."

Buck said, "Peter, winter's coming and I'd like you to make the decision on when you want to make the sale final and when you might want to move out. I'm not in a hurry. It's up to you."

"I talked with my sister, Mildred Denwick, last week and she says to come anytime I want. She has a bedroom waiting for me. It's short notice, Buck, but can you have the money and paperwork by next week. I can meet you at her house in town."

"Yes, I'll be there with the check and you can have all the time you need to move out. Let me know when you're through. It's been great seeing you again, Peter. I'll be leaving now for town."

The two men shook hands again and Buck departed, breathing heavily as he rode Titus toward Shadowcreek. It was a lot of money and a momentous decision. He was thankful he had had the foresight to save most of his money during the years prior to the Forest Service. Soon he would have a personal residence and would be able to ask Edna a question he had wanted to for some time. He wasn't going to do anything until he was the legal owner and had officially claimed the place he would call Singing Ranch. He also wanted to break the news to Edna at the proper time and place. It had to be perfect.

Edna left a note on her door telling Buck she was visiting Alice in Harmony for the weekend. He was welcome to come. Without hesitating, Buck and Titus traveled the ten miles without stopping. Leaving his bay at the livery stable, Buck walked to Alice's, a one story brown Victorian designed house with an enclosed porch on two sides and an oval window in the front door.

He knocked. Alice O'Neil answered and graciously asked him in. Buck and Edna hugged. They all sat down at the dining room table. It was apparent to Buck he had interrupted a serious discussion between the two ladies and he felt somewhat uncomfortable. Edna sensed his feelings and asked Alice if it was all right to tell Buck what they had been saying.

Alice said, "It's fine with me Edna. Maybe Buck will give me some good advice."

"Buck, you remember Sanford Picton. Well, he's written Alice a long letter telling her she has been in his thoughts and prayers for the past two years. He says he knows she would never move to a large city but wants to know if she would consider another small town if he became either a forest supervisor or quit the Forest Service, bought a ranch and settled down as a cattleman. He went on to state he still loves her, has stopped drinking completely and has taken control of his life. However, it would all be meaningless if she couldn't or wouldn't share it with him. He says a great deal more but that's the gist of it, isn't it, Alice?"

"Yes, you've covered it."

"Well Buck, do you have anything to say?" asked Edna.

Chapter Six - 1910

"Now, that's some letter. I knew when Sanford boarded the train two years ago he was crushed, but I also knew that he had a spark in him that wouldn't give up if there was any chance he could persuade you, Alice, to change your mind. In spite of the problem I had with him and his drinking, I really liked the guy. I spent several weeks with him almost 24 hours a day and got to know his so-called positive side. Since he has stopped drinking, I know he would make a good supervisor or even a good rancher. However, I'm not getting involved with you two on helping determine if he would make a good husband. I would think the first question would be do both parties love each other?"

Alice responded, "Yes, I know loving someone is of primary importance. I've gone through this before. I'm not sure of my feeling for Sanford. He wasn't here that long and I didn't get to know him well enough to know my true thoughts. He rather pressured me at the last and I had to say no. I really felt badly about it and called Edna for support. She and I have been discussing Sanford since he went away. Every now and then he would drop me a line and I would answer but until now he never seemed really serious about our relationship. I want to answer him but I want to do what's right."

"I've been giving her some ideas Buck, but Alice knows she'll have to make the final decision. She has several options about seeing him or not."

More discussion took place until Buck finally said they should all go to dinner at the Golden Restaurant. He planned to stay both nights in town and see Kent on Monday morning.

The three friends had an excellent meal, where the discussion related mostly to Sanford. Buck and Edna were very supportive of Alice in her decision dilemma. By the meal's dessert, she had decided to either meet Sanford at a neutral town or invite him to Harmony for a few days. With that they toasted each other with a glass of sherry.

At church Sunday morning Edna told Buck she needed to stay with Alice the rest of the day and would see him on his way through Shadowcreek during the week. He reluctantly agreed and walked to the office to work on some maps and reports. That evening he ate alone.

Buck was at the office when Lucy arrived and unlocked the front door. She motioned the ranger to her desk and proudly showed him her brand new adding machine. Up to that time she had been using mental addition and subtraction. Buck told her she deserved it and when he looked at the manufacturer's label he let out a little laugh. It said Dalton Company.

Buck decided to have some fun and told Lucy his plan. She hesitated at first but finally went along with it. When Kent arrived, he was as always, glad to see his Lakefield District ranger.

After discussing several topics and bringing the supervisor up to date, Buck said, "Kent, I sure want to thank you for buying me that adding machine. We can really use it."

"What the heck are you talking about? That machine is for Lucy. She got it just the other day."

"I believe you are mistaken. See for yourself. It says Dalton on the label. It's obviously made for the Dalton Guard Station and I'm taking it with me."

"Over my dead body. You're crazy, Buck. I'm calling Lucy in here to have this out right now." With that he asked Lucy to come in.

"Do you know what Buck has been saying about your adding machine?" asked Kent.

"Yes, I sure do, Mr. Bolton. He threatened and coerced me into giving it to him. Guess I'll have to go back to the long way again." Lucy was almost crying. Her voice cracked as she acted her part.

"The hell you will," roared Kent. "That machine stays in this office. Buck, you can't really believe the label meant the location of where it goes."

"You mean you're not going to let me take the machine?" replied Buck.

"You got that right. I thought you were more intelligent than that."

By this time the two plotters couldn't hold back anymore so they both burst out laughing. They even gave each other a little hug.

Kent sat there dumbfounded. Finally he said, "I know Buck put you up to this Lucy, so I'll forgive you. But next time there are any shenanigans taking place in this office it will be you and me against you know who. Understood?"

"Sure do, Mr. Bolton. You two are my dearest friends and next time it'll be you and me versus Buck." She left the room with a little skip and closed the door.

"Buck, you'll be the death of me yet," said Kent. "I've got to hand it to you, though. You really had me going. Better watch your behind. I might be there someday."

"I will, Kent. Thanks for being such a good sport. I couldn't let the opportunity go by. Guess I'll be going, if you don't have anything else."

"Wait a second. I received this Friday. It's a hilarious copy from the

Chapter Six - 1910

diary of a ranger my Forest Service friend sent me from the Arizona territory. It's dated August 15, 1900. He thought I'd get a kick out of it."

The Survey

Homested clame of Bud Brown, Bonefido squatter. This survey was run and plated on a variation of 9 degrez and 75 minits east of Polarus (or some other point I fergit which.) Wether looks like rane.

This tract is situawate in an unsurveyd terytory whicht when survyed wil probably be in Town ship 82 west of Range 3 north of grene witch.

That being no established corner in this basinety I build a pile of stones 4 fet high for a forrist reserves Monument, from which a miskeete tre bears north 7 degrez and 76 Minits east, a big mal pie rock bears west 27 degrez south.

Thense I run east 20 degrez north 48 chains and set corner no 2 a mal pie rock set in the ground (lots of other rocks around but this one has blubbers onit). Frun whicht a bald faced Cow with a little calf bears east 22 degrez south and a big steer going the other ways bears west 11 degrez north as other objext near.

Here I back site on Corner no. 1 and find that the variation had changed so I precede on a Tru Line.

Thense I run north 10 degrez west thru oke brush 21 chains to deep wash (here my dig got after a Mavric bull so I quit the survy and follo my Dog).

August 16, 1906. i start where i quit the survy yesterday and at 45 chains i set corner No 3 whicht is a oke stick set 1 ft in ground, whense a oke bush bears east, and the left hand and of a big cloud bears a little south of strate up, no uther Objext near.

Thense i run west 10 degrez south 15 chains or a little over to a high cliff to mark my line, when a white tale Buck jumped out of the oke brush and I kilt him with my sixshuter, (there i quit the survy and packed the mete to Camp).

August 18, 1906 i resume this line at the foot of the high clif wher my rock lit, i estermate the distance to be a little under 5 chains to the top so i allow i am now 20 chains frum corner No 4 whicht is a oke stick set in a dager wead, whense a smoke frum a forris Fier bears west 46 degrez north about 10 miles, no uther objext near.

Thense i run south 20 degrez east 11 chains as 15 steps to foot Of high clif i cant asend, so i shoot a spot on a rock on top to Mark my line, i climb the clif at anuther place an resum my Line, i estermate the distance to be about 5 chains a little back of strate down, so i allow i am now 16 chains and 15 steps from corner No.4. (here Bud Brown got a blister on his heel and quit chaneing, so) i continue on a tru line 735 yards as i step to corner No.1, whicht ort to be the place of beginning, but aint, so i allow thers some thing out of plum an adjust to my left and tie into the Corect corner, and the Place of beginning, contaneing 160 acres be the same more or less.

Bill Caltute, Forrist Ranger

All the while Buck was reading he was smiling and giggling. Afterwards he asked Kent if he was going to bring it to the next ranger meeting. Kent said probably not since the best part was actually seeing the diary as it was written.

It was early Monday afternoon and Buck was on his way to Shadowcreek. He stopped to see Elroy and check on Stub. They were both doing well. He saw Edna at the post office for a few moments and left for Dalton Guard Station—without the adding machine.

Snow was on the ground. The phone line from Lakefield to the outside world was already out of order. Buck was determined to make it maintenance free somehow, someday. There must be a better way, he thought as he rode north.

He stopped at several cattle ranches before Dalton. The allotment system seemed to be working although he did have to be firm a couple of times on denying requests to graze some animals before the season started in the spring.

Buck and Marshall worked on several maintenance items that needed attention at the station. Marshall had constructed a separate fire cache building and gave the property inventory to the ranger. Both men agreed it had been one heck of a fire season they never wanted to experience again. They split the forest patrol for the winter with Buck working on the west half and Marshall taking the east half. It would be 1911 before they saw each other again.

Before the week was over, Buck returned to Harmony, obtained the $1,500 and went to Mildred Denwick's house. That same day Peter and he signed the necessary papers and Singing Ranch was his. Peter said he would be completely out of the house by New Years. Buck thought that's when he'd do it. He could hardly wait.

1911

Forest Ranger's Song

What do you know in your dim proud cities
Of the world God made when God was young?
Have you ever lain by the limbs of nature
Or slept to the songs she has made and sung?

And Time is ours in the forest olden,
Time to listen and time to dream;
And Time to smile to each bird that flutters,
And Time to talk to each tumbling stream.

For we've given our hearts to the ancient forest,
To the stalwart pines and the sweatheart flowers,
To the winds that sing and the showers that sweeten
The marching months and the hurrying hours.

Evening comes; and a glowing campfire,
Wind in the branches sighs and sings,
Stars on guard and the night for cover—
Mine is a couch too good for kings.

W. P. Lawson

CHAPTER SEVEN

1911

Buck was nervous as cold water dropped on a hot iron spider. It was New Year's Day. Edna and he had celebrated the previous evening. He couldn't sleep so he got up, ate a light breakfast and walked to the city park to wait at the gazebo until Edna arrived. They had agreed to meet there after Buck said he wanted to talk. Edna was anxious also, since she knew Buck wasn't acting his normal self. She had no idea of what was about to take place.

It seemed hours to the ranger, although Edna showed up at the agreed time. As soon as she arrived, he asked her to sit on the bench in the gazebo. Buck immediately got down on one knee and looked straight into her eyes. They were those same beautiful deep blue ones that he had first seen just the other side of the post office counter. He had to move fast or he'd lose his nerve.

Buck stammered, "Eddy, I'm in love with you. I know I can't offer you much. I'm asking you to marry me. Will you be my wife?"

Edna looked at him adoringly. "Of course I will, Buck. What took you so long? I've loved you from the beginning of our friendship."

"What! Why didn't you say something. I had no idea. I just thought . . ." He didn't complete the sentence because he didn't know what to say.

"You silly, I would have never told you about my feelings unless you had expressed yours first. If I had, you would have hightailed it out of Shadowcreek and that would have been it. There were several times I thought you might leave but you didn't. I never questioned your reasons for doing things nor made demands on you. You would have panicked and felt trapped. That's why I never told you my true feelings, Buck."

"You're definitely the smartest woman I've ever met. I didn't ask before 'cause it wouldn't have been right to drag you to Lakefield station with no neighbors, electricity or shopping. But now I have a plan that should work so you won't have to go through any of that. Don't you think I was right to wait?"

Edna replied, "Buck, you still don't get it. If you had asked, I would gladly have gone with you to Lakefield in spite of its primitive living. We would have been husband and wife and together we would have made it work. I knew you wouldn't ask me to go to the station but I was willing to wait."

Buck was incredulous. "What do you mean you knew I wouldn't ask you to go to Lakefield?"

"Well, just say that Faye Wadsworth and I are good friends."

Buck now knew. He remembered his conversation with Faye when she asked about the two of them getting married. That was a long time ago. He should have kept his mouth shut.

"OK, let's have no secrets from now on," said Buck.

"That's a deal. Let's seal it with a kiss." And they did.

"So what is this plan of yours?" inquired Edna.

"Well, I've purchased Peter Blodgett's ranch along the Crescent River. He just removed his last belongings and is now living with his sister in Harmony. We have a permanent residence with phone, electricity and road into town, plus some land if we want to grow something and an area large enough for horses, mules, dogs, chickens and small animals. I could actually administer the district there better in winter. In the summer I could stay at Lakefield station and come back on weekends. What do you think?"

"I think it's a wonderful idea except for one thing."

"What's that?"

"Your statement about being at Lakefield during the summer without me. That is totally unacceptable."

"I just thought you would like it better to be near some neighbors."

"Buck, you aren't listening. After we're married I'm with you, no matter where you go or where you live. We are a package deal and the Forest Service better realize it."

"What about your job at the post office?"

"I'll send in my resignation."

"You are a wonderful, stubborn woman, Eddy."

"Sometimes I have to be stubborn around you, Buck."

* * *

News of Buck's and Edna's engagement traveled like wildfire throughout both towns, outlying ranches and Forest Service employees. Most of the women told Buck congratulations but added, "Why did you wait so long?" Most of the men gave their congratulations but added, "Sure

Chapter Seven - 1911

hope you know what you're doing even though Edna is a great lady." He shared their comments with Eddy and both agreed it was fun to be a topic of conversation.

They went shopping for an engagement ring. The wedding would be sometime in the spring before fire season started.

Buck was fine with getting married by a justice of the peace but Edna would have none of it. She said it didn't have to be a large wedding but she wanted the minister of the Shadowcreek Community Church, Jonah Stratford, to perform the ceremony. Buck agreed.

It snowed more than four feet during January. It was futile to work at Lakefield Station, so Buck did what he could from Singing Ranch. He fixed the barn and corral for Titus and Stub and brought in plenty of feed. The house was practically void of furniture. Peter had left a table, two chairs, bed and all the kitchen appliances and utensils. His sister had the same items at her place. Buck gave Peter an extra $20 for the furniture and two cords of wood.

Buck and Edna planned their future. She would give notice to the post office and move to the ranch after they were married. Eventually they would buy all new furniture, curtains and rugs. Edna would rent her house in town as a furnished dwelling. The extra money would come in handy and help pay for all the Singing Ranch improvements. The ranch house needed a good coat of paint inside and out. Looking over several dozen color combinations, they settled on a cream for the outside with a light green trim. The inside would be yellow in the kitchen, with wallpaper in the living room, dining room and both bedrooms. The braided rugs would be acquired after seeing how the walls and new furniture blended. Each decision was a new adventure for the lovebirds. Calling each other affectionate names came easily and spontaneously.

* * *

Due to the disastrous 1910 fires throughout the West, the district forester implemented an active program for development of a fire control and reporting system. He also established competitive prizes for best fire records on the forests. Fire reports were still being made on the 874 notebook forms. More emphasis was placed on complete reports for all fires. No clear definition of a statistical report had yet been established, so all false alarms were reported as fires, as were fires that were out before found. Causes of fires had not yet been formally classified. Breaking man-caused fires into incendiary and other categories gave the only significant figures. Too many fires that had been thought safe broke out again and had to be re-named. He said this was inexcusable and from then on fires must be out before being left alone.

Kent sent out several circulars to the rangers. Some of the topics were a formula for strychnine poisoning in range and pre-seeding areas for the control of rodents. Another spelled out the policy for protection of game on the national forests. It meant the Forest Service would from then on assist the states in enforcement of game laws. Rangers would be given state warden appointments. Another one told each forest officer he was not to accept meals or supplies from permittees or residents of the forest. The following quote was included. "The inborn hospitality of the forest user often makes him unwilling to accept pay when you eat at his ranch or camp, or get fresh meat from him. You must insist on paying."

* * *

Near the end of April, Buck received a call from Hugh Tanner at Barnesville stating the Reliance Lumber Company would complete the Split Rock Timber Sale in three days. Buck was welcome to be part of the final inspection. He left the next day on Titus with plans to swing around to a couple of his own timber sales on the north side of the district.

The final inspection was led by Hugh and Collier Newton, president of Reliance Lumber Company. Buck observed and learned. Except for a few minor corrections on slash and cleanup, the two parties agreed the sale was successfully completed.

The next morning Buck thanked Hugh for the invitation and proceeded south and east along the trail that would take him near the timberline on Tip Top Mountain.

The day was clear. The snow had melted and Buck was thinking about the wedding and Edna, as Titus ambled along the vastly improved trail he first rode in 1905. Vibrations from the mountain made him stop and listen. He knew the Indians considered the mountain sacred but he didn't think it was haunted, which was how it sounded.

The volume increased as he rode. Leaving the trail to follow the sound, Buck came to the edge of the tree line and looked north to the apex of the bald mountaintop. He was amazed and fascinated by the sight. For the first time in his six years on the district he had stumbled on an Indian ritual. A fire was in the center, surrounded by a ring of men and women in costumes. The drums beat a steady rhythm as the dancers hopped, skipped and jumped in one direction and then the other. It had

Chapter Seven - 1911

to be the Hondo Indians from the reservation adjacent to the north boundary of the Lakefield District. The costumes contained feathers from eagles with red feathers bound on the base of a band around the head made from red headed woodpeckers. They wore deerskin-like chaps. The women had no upper garment to cover themselves but wore a skirt with tinklers around the bottom that made noise as they danced.

Buck was close enough now so he could see Olivella shell necklaces some women wore. He knew they were used for money but the most valuable necklaces were the Hudson Bay trade beads. They were handed from mother to daughter. He noted that several children were playing a stick game away from the ceremony.

Buck was thankful he had spent time at the Harmony library studying the beliefs and customs of the Hondo Indians. He figured someday there would be an encounter between an Indian of the tribe and himself. They thought highly of the wiley coyote and this looked like some kind of a coyote related ritual. At other times he had come across several family burial plots in the forest but had left them alone. He never mentioned them nor did he mark them on any map. It was better that way, he reasoned, so grave diggers and trophy hunters couldn't desecrate the sites, as he had experienced the previous year.

Buck continued to watch for more than an hour. He stayed stationary and quiet. The Indians took no notice of him but he realized they knew he was watching. Finally, he turned and entered the forest with the singing, yelling and drum beat ringing in his ears. He would try to describe the scene to Edna but knew he couldn't do it justice.

* * *

Buck and Edna set their wedding date for the first Saturday in June. They sent out invitations to their Forest Service and townspeople friends. Buck thought it prudent not to send any to his favorite ranchers or folks doing business with the Service. There could easily be hard feelings from someone left out. They wrote the vows together instead of using the traditional words. It would be a simple ceremony, with brother Alva as best man and Alice O'Neil as matron of honor. The Reverend Jonah Stratford would officiate at the Shadowcreek Community Church.

Buck figured only the folks on his district plus the supervisor's office would attend. He was wrong. Kent called a two day ranger meeting at Harmony for the Thursday and Friday before the wedding. When Buck found out, he had mixed emotions and asked his boss if he thought the other rangers might feel obligated to attend. Kent just smiled and said it was all set and he wasn't going to change it.

As the fateful day grew closer, Edna and Buck painted and papered the house at Singing Ranch after working hours. She was successful in renting her home. The new tenant agreed to move in on the Monday after the wedding. They had a great time picking out furniture, beds and bedding, rugs, curtains and wall hangings. Everything was ready for the Stonewalls to spend their first sleep-in on the first Saturday night of June.

* * *

Luke and Marshall were invited to the ranger meeting at the Harmony Community Hall, along with the five rangers, plus Kent and Stuart. Before the meeting started after lunch on Thursday, Buck received all sorts of barbs and negative comments from his peers. They needled him mercilessly. He took it in the spirit it was intended. Kent called the meeting to order. First thing he did was look at Buck, shake his head and call for a minute of silence. Laughter ensued instead of silence.

Kent and Stuart went over the budget and expenses for the current fiscal year and the one starting July 1st. Each district ranger justified his work and expenses. The new projects were listed and prioritized. There was money for Buck to build a decent lookout somewhere on his district. The other districts were to build trails, campgrounds, bridges, weather stations, make timber sales, plant trees, collect seeds and several other additional construction projects. Kent mentioned the Pinedale nursery would furnish the planting stock. The current year planting would be experimental. He gave each district the number of acres they were to plant.

It was past the time to adjourn, so Kent asked if anyone might want to play a friendly game of poker at the hall after supper. All hands except Buck's were raised and Stuart was assigned the job of coordinator.

Buck and Edna knew that Thursday's supper would be their last as single people. Edna had left for Harmony after closing the post office. Friday night was reserved for ranger stories. Buck had promised he wouldn't take a drink. He surely didn't want to mess up the ceremony or forget his vows. Edna trusted him but asked what would happen if he didn't attend story night at all. He answered that he would have to write unofficial letters of apology to each ranger asking for forgiveness. It was an unwritten rule. However, he would leave after the first round of stories. Edna understood.

Chapter Seven - 1911

They chose a dark corner table in the Golden Restaurant. Ignoring everyone else, they reminisced about their past six years together. They laughed, held hands, looked into each other's eyes, drank wine, talked about the future and touched knees and feet under the table. The other rangers left them alone and ate as a group. Buck and Edna walked in the main city park that evening.

Next morning when Buck arrived, he noticed a somber mood from his peers. It wasn't that they didn't tease him. It was something about their demure demeanor. Catching bits and pieces of the conversation, he learned that there was one big winner the night before and six losers. Tim, Ralph, Alva, Hugh, Luke and Marshall had all lost to Stuart in draw and stud poker. They were beaten badly.

Buck found out later from Kent, who didn't play, that the deputy supervisor had been a professional gambler in his earlier years. He didn't cheat because he didn't need to. The others were no match for his prowess. After the event, however, the rangers took more of a respectful attitude and admiration toward the deputy. Before that he had been looked upon as a cold, reserved and formal type individual.

The meeting came to order and Kent stated there was a large beetle control program going to start the following year along Clear Creek in the Humbug drainage of the Canyon Springs District. He wanted at least two more rangers to help Tim with the work. A staff of specialists would be established by the district forester to assist the Maahcooatche in surveys and control. The Bureau of Entomology would furnish the scientific information needed. Both Buck and Alva said they would help.

Kent handed out a proposed procedure from the district forester for counting stock at the start of grazing season. He wanted comments from the rangers. After reading it, Hugh raised his hand and said, "Asking the stockmen for an exact date in advance of entrance would not work because gathering for the drive is hit or miss and they can't tell how long it will take, so two or three days notice is all we can expect." The others concurred.

Basic questions on organization, priority of work, patrol versus lookouts, fire investigation and many others were discussed.

Before adjourning the meeting, Kent said there was one more thing he wanted to announce. His countenance changed as he said it. One could have easily heard the proverbial pin drop.

Kent spoke. "Men, what I am about to say is probably the hardest thing I've had to say for the past eight years."

The rangers looked worried.

Kent went on. "I've been your supervisor since the days of the Department of Interior. You have become my friends as fellow employees of the Forest Service. You are the hardest working, most fun loving, responsible, clear thinking damn bunch of cowboys I've ever known or ever expect to know. What a magnificent group. Life goes on and so does the Forest Service. I've been asked to transfer to a forest having major problems. It is a promotion but that isn't the reason I'm leaving. I report to my new supervisor's job next month. I already informed Stuart and Lucy."

A loud "Noooooooooooooooo" echoed through the room. Once silence reigned, Buck stood and made a short speech about what Kent meant to him and, undoubtedly, the others. At the end there were three cheers for a teary eyed supervisor.

It was agreed that story time would begin in three hours under a secluded group of trees at the city park. It would be critical that drinking didn't get out of hand inside the city limits. Each ranger could bring enough for three or four personal shots. That was it.

Buck ate with the group, since Edna had left Alice's that morning for Shadowcreek to prepare for the wedding.

Kent was designated to start the stories since the reasoning was they were at the headquarters town.

He spoke. "This is not a funny story but it is true. About a month ago I was walking home and saw an old farmer on a wagon with a team of horses moseying along the main street of town. All of a sudden, something spooked the horses and they took off in spite of the valiant efforts of the driver. The problem was the floor boards of the wagon weren't nailed down. They started moving and finally the farmer fell under the wagon holding on to part of the frame. Two guys ran from a store and tried to catch the animals but missed by about three feet. The farmer was hanging on to the lines and was between the front wheels and the hind wheels. When the railroad tracks loomed ahead, the horses made a sharp turn. That's when the driver let go. I ran to his aid. He had all the flesh torn off one side. After taking him to the hospital he told me later it took some time to get the gravel out."

Kent told the group the farmer had healed but had learned his lesson about loose floor boards.

No one took a drink after the story. It was a combination of Kent's previous admonishment to behave and the subject matter of the tale.

Buck told the next one, "I heard about a ranger who was packing feed for his animals on a bulky mule. He, his horse and the mule needed

Chapter Seven - 1911

to cross a narrow bridge over a raging river. As the mule made the halfway point, he started shaking. His load was unbalanced and he became frightened. Finally, he fell into the frothy liquid. This meant the load increased doubly, due to water saturation. The animal soon disappeared and was never found.

"That same year, when property inventory was due, the district came up short on some items. The ranger sent a letter of explanation, stating the property went into the water with the mule. The supervisor's office accepted the response. The following year the same thing happened. More property was missing and the answer was it had to be lost on the mule. A third year the ranger tried the same tactic. This time a harsh letter from the supervisor said, 'For the past three years you have claimed property was lost due to it being loaded on the drowned mule. You certainly lost a wonder animal. It must have been the largest mule ever to exist. We added up all the weight of the lost property and it came to more than 2,000 pounds. Next time please try another excuse. Maybe you have a giant bear somewhere on the district which can take the blame.'"

The rangers laughed with approval and the first round went down painlessly.

Tim was next. "It seems that a ranger on the forest to the south of me was out one day in winter and happened to come across a bear still in the torpor of hibernation, sprawled under the protection of a large pile of downed trees. At the time, the ranger was painting some fences, so reacting with a devilish impulse, he took the brush and painted a white stripe right down the middle of the bear's back. That didn't end it. When spring came, a group of hikers ran to the ranger station with mouths agape and eyes wide open. Talking all at once to the ranger, they said there was a huge striped skunk that they sighted about a mile back up the trail. They figured it was some kind of mutation. Since skunks were carriers of rabies they thought the ranger would want to hunt it down. He listened intently without giving away the facts. He said others had reported the huge skunk but assured them that light rays and distances often played tricks on a person's eyes in the woods. There was nothing to worry about. Often as he rode through the forest he had imagined little elves, barking bears, 30 foot snakes and giant deer. Most of the group bought it but one little girl was skeptical as the group hurriedly left for civilization."

A second round of drinks was forthcoming.

Ralph spoke. "I'm not sure I should tell all you blabbermouths this next story 'cause it's about me and what happened last winter."

The whole gang said he should tell it and they swore themselves to secrecy.

"Yea! And I have some swampland I'll sell cheap," he replied. "Anyway, knowing you guys can't keep your traps shut, I'll proceed. I was heading for Fulton Station on a brand new horse. I found myself in a blinding snowstorm and darkness was coming on. Along with a high wind, it became a blizzard and I could see nothing. I was becoming exhausted and badly chilled. The horse was tiring. There was no alternative but to keep going. The snowdrifts were extremely deep. Suddenly we were within a few feet of a building. It was small but large enough for my horse and me. Luckily the door didn't have a lock. What a relief to be in out of that wind and blinding snow. The building was vacant. There was no indication of to whom it belonged. After spending a sleepless night shivering, my clothes were still damp in the morning. The wind and falling snow had ceased. I looked outside and there was the Fulton Ranger Station. My guard had just completed our tack room and I had never seen it on the inside. So, I'll admit, by my own stupidity, I had been within 100 yards of my own house and a warm, dry bed."

At this last piece of news the whole group went into hysterics. There was nothing better than a story from and about one of their own. After more taunting and heckling, they ended by praising Ralph for being a good sport.

It was Alva's turn. "This is also a true story about a miner I was having trouble with, but no more. It seems this miner had about 50 pounds of dynamite in his cabin for a couple of years. I came by one day and saw it. Due to safety, I asked that he either place the dynamite in a powder house or dispose of it. Since he didn't have a powder house, he chose the latter. I thought he would bury it but after carrying it to a meadow a couple of hundred feet, he returned and grabbed his trusty 30-30. I told him not to pull the trigger but he did anyway. The boom was terrific. The concussion knocked all the dishes off the shelves. The stovepipe was blown off the roof and the miner who hadn't taken cover was laying flat on his back about five feet from where he originally stood. Luckily, I was standing behind a corner of the cabin. All the glass in the two windows was out and there was a hole in the ground where the powder had been, big enough to bury a horse and then some. I

Chapter Seven - 1911

helped put the stovepipe back and replace the unbroken dishes. A week later I saw him in town with some new panes of glass and some putty. I never told anyone on the district or in town what had happened and I'm not going to mention his name here. However, we are now friends and he hasn't given me any more problems."

Since the story was true but could have ended in tragedy, there weren't any comments, except they all agreed it was a good tale, especially with Alva's animation. He performed almost as well as his older brother.

Finally it was Hugh's turn. "Since the previous two were first person true stories, I'll continue the tradition. Last year I took a man to the local justice of the peace and charged him with leaving a campfire burning. The judge asked the defendant if he was guilty or not guilty. His reply was, 'Not guilty!' The judge said, 'Not guilty? Of course you're guilty! You've owed me ten dollars for the past five years. Since the fine is normally fifteen dollars, I fine you twenty-five dollars.'"

The first round of tales by the rangers were completed until someone said, "Hey, what about Stuart?"

Stuart had been sitting off to the side, taking in the presentations and enjoying them like everyone else. He said he had no Forest Service ones to tell but he did have a story about his gambling years. This got everyone's attention. "It seems that when I was quite young, I carried a great deal of cash, which was a bum idea. I joined a bunch of strangers around a card table one evening and feigned ignorance of the game of poker. You should have seen the smiles. As we played, I managed to win a few hands out of blind luck from what I told the others. I always stayed with an excellent hand but threw in with anywhere from a nothing hand to a good one. I never varied my system nor bluffed for several hours. They learned I was predictable. I lost but not too much. By 4 a.m. there were only four of us left. The others had dropped out. The game was no limit. It was agreed there would be one last hand before calling it quits. The game was draw poker, with each card turned one at a time with bets placed each round. The pots in this type of game usually grew to tremendous amounts. We all anted up and the cards were dealt. I had an ace, a king, a queen, a jack and a deuce. The odds were long against picking up a ten so I said, 'No cards!' This caused a look of concern from the other three, who had asked for one, two and three cards. I laid my cards face down, so the ace was first and the deuce last. With my ace I went with a daring bet of fifty dollars. Everyone stayed in and the fourth person doubled the fifty. It went around again and the second card was turned. The person to my left had a pair, so he bet first. In went two hundred dollars. When it got to my turn, I doubled it. After that round we all turned over the third card. The same person started with the same pair and bet five hundred dollars. I again doubled it. When it came my turn again, I raised it to a thousand dollars. Everyone stayed in. The pile now contained more than $12,000. The fourth card was turned. When the others saw my ace, king, queen and jack, they all dropped out. The money was mine. They knew in their own hearts I wasn't bluffing, when in fact I was. So you see men, don't take anything for granted. I didn't show the others my last card 'cause I didn't have to, according to the rules. But on the other hand, I didn't rub it in. The money ended up safely in the bank the next day."

Stuart received an ovation of admiration. The group was behaving well, compared to other sessions. Kent started them off with a second and third round of stories. The meeting broke up by 11 p.m.

* * *

The big day came. By tradition Edna wasn't allowed to see Buck, so he struggled alone with his tie and finally got it set with a semblance of satisfaction. The church was filled and the procession and music started on time. Edna was beautiful in her ivory, organdy dress, trimmed with lace. She wore a garland of tiny flowers in her hair. Buck was handsome in his only suit. Although the loving pair didn't falter it was obvious they were nervous but happy. Their vows were made with adoration, feelings and promises, with a reverence for God. Afterwards everyone said it was a beautiful ceremony. Refreshments were consumed and Elroy had a horse and wagon ready with streamers and metal noise makers attached. Edna threw the bridal bouquet which Alice caught. With good wishes, kisses, hugs and waves, Mr. and Mrs. Buck Stonewall sat together on the wagon seat and were gone. Between town and Singing Ranch they laughed and talked.

That evening, as the dying rays of the sun transferred the light to a full moon they climbed into bed. At midnight a horrendous clamor broke out. Yelling, pots and pans clanging, bells tolling, whistles shrieking and banging on the front and rear doors brought Buck and Edna bolt upright in bed. Groggy from sleep, they stumbled about as they dressed hurriedly.

"Oh Lord, it's a shivaree,"[24] exclaimed Buck. "Now I know why Kent smiled when I asked him why he wanted the ranger meeting a day before our wedding."

Chapter Seven - 1911

"The noise is deafening," yelled Edna.

Buck barely heard her. On opening the front door, a dozen men poured in. They were all laughing and joking and demanding the newly-weds bring out the food and drink. The person having the most fun was Kent. He kept saying to the ranger, "Got you!"

"You sure did, boss," said Buck. "This makes us even. Right?"

"No, I'd say it was still about 20 for you and one for me," laughed the supervisor.

Edna scurried around, gathering containers for drinking and plates for eating. Luckily, they had stocked up with several weeks of food. Since there weren't enough chairs, several sat on the floor. After eating, the invaders demanded Buck bring out his harmonica and Edna sing. The duet continued for a good half hour and finally Kent informed everyone it was time to leave. He told the Stonewalls they were good sports and their wedding night would be memorable for at least two reasons. Falling into bed for the second time that night, they realized it was short sheeted and sprinkled liberally with rice.

* * *

True to her word, Edna packed and moved to Lakefield Ranger Station. She made several trips on her own in order to make the place homier. Buck was away most days and often at nights. She didn't mind. It was better than seeing him once or twice a month. One of the ranger's motivation for returning to the station as often as possible was the anticipation of eating a delicious home cooked meal.

During the July 4th holidays, all the rangers, assistant rangers and wives attended a going away party for Kent. The public was invited and more than 100 people attended. Stories were told and Buck embellished on some of the outlandish events that he and the supervisor were involved with. Included were references to Kent becoming soft by driving his Model T everywhere instead of riding Fiddler. He said he admired Kent as much as any man he had ever known and that he had been his best boss by far. He thanked the supervisor for putting up with his shenanigans and wished him well in the future. Others spoke and presents were bountiful. Kent addressed the room but had to cut his speech short as he choked up.

* * *

One morning as Buck was busily working in the barn at Lakefield, he heard the steady beat of hoofs coming up the trail. He recognized

Stuart Brewer immediately but the other one was a stranger. Upon dismounting, Stuart introduced his companion as Cliff Clayton the new forest supervisor.

Cliff said, "So, you're the infamous Buck Stonewall that Kent told me about. I'm very glad to meet you and hope we'll have a great relationship. I'm looking forward to working with you."

Cliff was almost six feet tall, with receding hair and a mustache, large ears and a little on the heavy side. He was in his mid-thirties and wore a wedding band.

Edna invited them in for refreshments. They toured the compound and Buck asked if he could show them more of the district. They explained it was only a day trip. Afterwards, Buck told Edna he thought Cliff seemed like a good man but was the most pleased with the rapport the two men shared. It would make work much easier to have a happy relationship between supervisor and deputy.

* * *

Every year since Buck and Luke had constructed the solid insulator telephone line from Lakefield to the Crescent River commercial line, the two men had to make numerous repairs each spring. It was one of those tasks accepted throughout the Forest Service.

Buck read about a ranger named William Daughs inventing a split tree insulator. The first one he whittled from a piece of Douglas fir bark. It was in two parts and allowed the line to ride free through the oval hole in the center. The Service recommended its use and at the same time adopted No.9 iron (galvanized wire), which was heavier and stronger than the formerly used No. 12. The No. 12 wire was used to bind the insulator halves together. This same wire was hung in trees by banding it into hooks at both ends that hung on a staple driven into the tree. By allowing the No.9 wire to ride freely when a tree fell across it, the line seldom broke. The slack wire could be pulled from either end.

Buck was excited, since it would mean much less maintenance and repair work. He petitioned the supervisor to let him replace the old No. 12 main line and the old solid insulators. The approval came through quickly. It was easily shown the economic benefits of such a change.

As in 1906, Buck ordered miles of wire and hundreds of insulators. The wire was cut into quarter mile sections. By early August, Buck and Luke had brought the merchandise to Lakefield. This time they had the luxury of a third man hired specifically for the job. Marshall was busy with other important work. Two would climb and the man on the ground would lay the wire on the ground and place one half of the insulator under it and one half over it and tie them together. The amount of slack was determined by standing on the ground between trees and reaching the line at midpoint. If a line crossed a trail, then the slack would have to be raised so that a man on horseback could reach up and pull it down at midpoint.

In later years the tie wire was wrapped around the tree at least twice to hold the insulator in place. Shims were used between the tie line and the trunk to prevent the wire from strangling the tree.

* * *

In late August Edna told Buck she had received a call from someone representing the Crescent River school board. The school was located about a half mile west of Singing Ranch. It was the same one room, one teacher school that Luke's children Matthew and Mark attended. She said last year's teacher had moved away after the contract expired and they hadn't been able to find a replacement. When they asked if she would consider teaching, Edna reminded them there was a rule about no married teachers. They knew that but would allow a waiver if she would sign on. School was about to begin and the children needed an experienced leader. She said she would discuss it with her husband and let them know immediately.

"Sweetheart, I'm really torn about this. I want to stay with you until winter but I also would feel terrible if all those children didn't continue with their education."

"Well, Eddy, it's only for one year until they get a permanent replacement. We should think of the kids first."

"No wonder I fell for you, Buck. Thanks for being so understanding. I'll let them know. School starts in only ten days."

"I'm going to ask Luke to sell us a horse so you can ride back and forth. You need one anyway, instead of renting from Elroy all the time."

By the time Edna's first day came, she owned Leo, another bay like Titus but not as large. There were 14 boys and girls ranging from the first through eighth grade. Edna loved it. She was back in her element. The school board and children were lucky to have her. For several years afterwards she continued as a substitute teacher each winter.

Luke, Marshall and Buck were busy fighting fires that fall. The largest was controlled at 35 acres. They hired men from timber sales and ranches. Buck would have to catch up during the winter.

After moving back to Singing Ranch from Lakefield, Buck spent more time in Harmony working with the new supervisor. Cliff had a Model T Ford, along with one or two others in town. Where would it all lead, thought Buck?

The Stonewalls celebrated Buck's 35th birthday on July 9th and Edna's 30th on October 15th. They became involved with the Shadowcreek Community Church and volunteered their time on many community projects, especially helping the handicapped and poor. Edna thought it was important they share their good fortune.

One morning they awoke to find someone had left a puppy in a box at the front door at Singing Ranch. It was whining and looked hungry. Buck agreed to let Edna keep it but made it clear he didn't have time to care for it, nor would he allow it to travel with him behind Titus after it grew to an adult. Edna named her Nellie. She thought it was a black Labrador retriever. Her golden retriever Molly tolerated Nellie at first and later they became real buddies.

Buck was gone for more than a week on a late season fire. Rain had finally put it out. Averaging two hours sleep every 24, he was exhausted and on entering the house told Edna he wanted to get some shuteye. Taking off his boots and pants, he fell into a deep and contented sleep. After an hour, Edna came into their bedroom and looked down at Buck for a long time, thinking what made the man she loved tick. What motivates and drives him to such extremes as a ranger, she wondered. It had to be for personal satisfaction, going all the way back to how he was brought up. She thought the Forest Service was lucky to have such an employee. She knew most of the rangers in the agency were hard working men of character but she believed Buck was someone special. It made her proud and happy to be a permanent part of his life. Quietly she walked away. Her cup overflowed with joy. If only time could stand still.

Chapter Seven - 1911

1912

The Night Trail

I rode on a lonely trail when night
    From the depths of the canyons drew
A dusky veil over crag and height
    And the wild land dimmed from view.
And I paused a space on the rock-strewn rise
    Where the trail to the canyon dips,
To watch how the day-flush leaves the skies
    Through the west, where a rim of mountains lies
With a fading flow on their tips.

In the moment's hush when the day was done
    And the still world seemed to wait,
An outcast coyote wailed alone
    And a far elk called his mate:
And it seemed that the wild things voiced a dread
    Of the gloom and the mystery,
Of the Sense of Fate that with silent tread
    Crept close around, and whose calling led
Into ways that they could not see.

And my horse goes true to the end of the trail,
    Where the light of the camp shines out—
And true goes our purpose that will not fail
    Till we pass through the gloom of doubt:
True goes the purpose that leads us still
    When our cause knows the hour of night
Knows the shadows of greed and of selfish will—
    For we know we but ride in the gloom until
Our way has an end of light.

                Scott Leavitt

CHAPTER EIGHT

1912

During the next few years there was rapid expansion of the use of automobiles on the Maahcooatche National Forest, even though only a fraction of the entire area was accessible by wheeled vehicles. Primary travel was still by horseback. Communication had been established to all major stations, which caused a reduction in the volume of letter writing.

The Service had finally received approval to pay mileage on privately owned cars and motorcycles. The following rates were listed:

5 cents per mile for two and three passenger vehicles
6 cents per mile for five passenger Fords—T Model
7 cents per mile for Detroiter, Paiges, Studebakers, Overlands and Buicks—light models
8 cents per mile for Reos, Carters and Stanley cars and heavy model Buicks, Studerbakers and Overlands.

A system of cost keeping for each vehicle was proposed in order to get reliable data on which to base mileage rates in the future. Claims were to be submitted monthly as an expense item.

Supervisor Clayton wrote to each ranger:

"All the men employed on the Maahcooatche National Forest will, I am sure, take great pride in the knowledge the forest has captured the district's prize shovel as a result of the fire record made during the past year.

"The shovel now in this office, painted green and with the words FIRE PROTECTION PRIZE attests to the fact that we have it. A shovel is the prize for being first, a hoe for second and an axe for third."

The district forester sent a letter of instructions for a cooperative study with the U.S. Weather Bureau to formulate a method to calculate the relationship of winds with severe forest fires. This was the beginning of forest fire research. Several special fire studies were undertaken. A complete analysis of the time factors in suppression of the year's fires

were requested. The instruction for the study divided the "life of a fire" into four periods:
1. Time of start to time of discovery.
2. Time of discovery to time of report received by officer or person equipped to start suppression action.
3. Time of report to time actual suppression action started.
4. Time from when work started to time of extinguishment of safe control.

Buck and Alva left their districts to help Tim on the beetle control program along Clear Creek in the Humbug drainage of the Canyon Springs District. Five people from the district forester's office were detailed to the $9,000 program. They camped several weeks in two four man tents, taking turns telling stories, playing cards and having a raucous time. The cook had his own tent. They moved camp twice during the project. Buck wasn't use to only an eight hour work day but the district folks insisted on it. He was glad to have brought some reading material.

The beetles traveled in broods. Trees that had been injured by fire and lightning were more likely to be attacked than sound, thrifty trees. They had no choice place for attacking the tree, sometimes near the ground, midway or on the extreme top. After the trees were felled, the men commenced burning the bark. A compass man and spotter checked up on the board feet in every tree felled and bark treated. Sometimes 100 or more attacked trees were found on a 640 acre section of land.

Near the end of the project Buck wrote the following poem:

> Come all ye people if you want to hear
> The story of the bug crew in the creek called Clear,
> Of a terrible country and long career
> For the Rangers and Bug Men far and near.
>
> We made our camp on a cold wet Sunday
> With our horses gruntin, full of oats and hay,
> We put up our tents by the candle ray,
> And we ate our supper when the dawn was gray.
>
> We built a fire in the middle of the tent.
> She ripped and roared and away she went;
> I tell you fellows that it ain't no joke
> When your bloomin' old tent gets full of smoke.

Chapter Eight - 1912

Now all ye people when ye spot a bug,
No matter if our crew is housed up snug,
Just tell us about it and we'll paste his mug,
And we'll join in the chorus while his grave is dug.

Chorus

We chopped 'em all down,
You can't find a beetle;
We bucked 'em all up,
Can't find a bug;
We burnt 'em up clean,
Can't find a beetle;
Oh you can't find a beetle on the Big Humbug.

* * *

During the spring the Bureau of Public Roads hired a survey crew to work on a proposed road along the Crescent River, between Long Tom and Luke's Crescent Ranch. A filed report recommended that Queens County cooperate with the U. S. Forest Service fifty-fifty on the cost of construction. The project started in the fall.

School was out and the Stonewalls moved to Lakefield Station for the good weather months. Buck was elated in that the phone worked with no repair necessary. He had Marshall work with the ranchers, getting their animals back on the range allotments. He had Luke pinpoint a good spot to build a lookout somewhere on Florin Peak. Although Tip Top Mountain was higher, he didn't want to create animosity with the Hondo Indians and build a structure in the middle of their sacred ceremony grounds. He thought it might be necessary someday but not right then.

Buck was given the plans and funds to build the first manned lookout building on the forest. Cliff asked if he could complete it that summer. In his usual can do way, Buck answered he could unless called away for too many fires. The drawings were quite detailed, so he could order the exact size and amount of lumber and material for the foundation.

Luke found the spot so the two men loaded a long horse and mule pack string with supplies and tools. It was easier to travel the Crescent River road until they turned north on the trail to the peak. In places, this trail had narrow hairpin turns and rock outcroppings. The animals had to be strung out because of the length of the lumber. It was slow going.

Several times either Buck or Luke dismounted near a tight turn and individually escorted each animal to the safety of a straight path. On arrival at the site, the animals were led into a makeshift corral that Luke had built earlier. The men would sleep on the ground.

At daybreak the work started with mixing concrete by hand for the foundation. Metal rods for reinforcement were used in each of the four concrete footings. Buck laid out the lumber and hardware to make an orderly construction sequence while Luke took the animals back to town, returning the next day. Marshall also showed up. Buck was glad to see him since it was a real struggle to erect the large vertical lumber stringers and cross arm supports with only two men.

The Florin Peak lookout was a one room, six by six cupola with windows without glass on all four sides. It was accessed by a ladder through a trap door. The lookout stayed in a tent at ground level, directly under the four struts secured on top of the concrete.

They had brought ladders, rope, pulleys, levels and numerous tools. After building a station, a bridge, a campground and several other structures over the years, Buck knew what was needed and how to do it.

One morning they heard the soft tinkle of a bell. After some moments, there emerged a procession that might have been common over 2,000 years ago. In front was a bearded man in flowing robes leading a docile donkey loaded with all manner of family possessions. He was followed by a young woman and two small children, also dressed in robes. The three builders stood and stared. They were still gawking when the man asked where the trail went.

Buck responded, "Well, sir, we're near an intersection, about one hundred yards in that direction," as he pointed east. "If you go east, you'll meet my wife Edna at the Lakefield Ranger Station. If you go south, you'll come to the Crescent River. There is no bridge at that point. North is to the town of Barnesville. West is a long route to Long Tom. Would you like a drink of water?"

"Yes, thank you very much," said the tall stranger. "Is there enough for all of us?"

"We have plenty," answered Buck.

After the group had their fill, they thanked the three men and headed east to the intersection.

When the lookout was completed Buck checked with Edna at Lakefield, Hugh at Barnesville and Alva at Long Tom. No one had seen the procession. It was as if they had vanished into the forest primeval.

Chapter Eight - 1912

The ranger thought about it many times but came up with no logical answer. He accepted it finally as a mystery straight from the Holy Land.

* * *

Nellie was almost fully grown. She would whimper and whine when Buck and Titus left the compound. One day she broke her restraints and dashed madly up the trail. She found them two miles from the station and Buck knew he'd have to take her with him. She did everything Buck requested and maintained her speed and stamina throughout the day. Many times after that, Edna laughed privately as she remembered Buck's statement about Nellie staying home. From then on, she never did. It was apparent that Molly was Edna's and Nellie was Buck's. Nellie was trained to be a hunter and retriever. She eventually won several ribbons at the county fair.

* * *

Buck was sitting in his favorite chair, staring out the window toward Crescent River. Suddenly he turned to Edna and said, "You know, dear, I'll have been the ranger at Lakefield District for eight years next April. We've never had a vacation together just as we didn't have a honeymoon. You haven't taken it easy either, since the day I met you in 1905. I think we deserve a break from responsibility for a week or so. We can go anywhere you like. I'm sure you've wanted to see certain things you haven't been able to before. What do you say, Eddy?"

Edna looked at Buck lovingly and smiled. "If there is anyone I know who deserves a break and some relaxation from all those hours and years of toil, it's you, Buck. You've earned it and yes we should do it sometime."

"What do you mean by sometime?"

"Well, honey, I saw Doc Cleary yesterday and he said, 'Edna, you and Buck are going to become a family. For three months you've been with child!'"

EPILOGUE

To My Old Comrades

Although I am tired and weary
I will take up my pen and write,
As I think of those days in the Service,
Those days so busy and bright.

Upon the screen of my mem'ry
There flashes the faces of men
With the real red blood of their fathers,
More used to the rifle than pen.

With a smile they faced all the dangers
Of storm, of flood and of field,
Nor were they e'er known to falter
Or an inch from plain duty to yield.

It may be I never shall see them again,
But my best wishes go with them thro' life,
And may they be happy and prosperous too,
Also good luck to the brave Ranger wife.

C. C. Hall

EPILOGUE

The days of the pioneer custodial rangers were numbered. Their ranks were being filled by college educated professional foresters. The utilitarian years of the Forest Service were approaching.

The future would change the system when employees had been expected to work twelve hours a day or travel without pay on off days as well as work days; when the lowest paid employee knew every other employee's name, plus those of the spouse and children. The old Forest Service consisted of non-specialists. Each one knew how to fight fires and each knew he was expected to participate in fire control.

There are, however, some basic tenets of the agency that have endured. It is still in the business of land management. It is still decentralized. Employees must still make many independent on the ground decisions. It is still in the business of timber sales, tree planting, grazing allotments, fisheries and wildlife habitat, fire management, recreation management, road and trail construction, preserving botanical and archaeological sites, law enforcement, search and rescue and a myriad of other functions.

Those first rangers were true American heroes. It would be extremely difficult for someone today to fit into their shoes and do the job that was required. The physical work alone was daunting: long hours, six to seven days a week; sleeping on the ground with maybe one blanket; working in all types of weather; searching for and rescuing lost people; experiencing loneliness; crossing rivers; fighting off hordes of insects; lifting logs and bodies; fighting fires for weeks at a time with 100% hand tools; going days with no food or weeks without clean clothes; walking for many miles with a full pack; scaling rocks; negotiating rapids and much more.

The knowledge needed to perform was astonishing. They were required to know and ride horses for many hours and miles at a time; do blacksmithing; shoot guns; be proficient in the use of hand tools such as axes, saws, sledge hammers and many more; load pack horses and mules; set up a compass and run a line on a prescribed course; survey; estimate timber volume; record tree size and species; fell trees, limb and pile without damaging others in the stand; build cabins, corrals and other structures; build and maintain trails; climb trees for phone lines and fire observation; enforce grazing, mining, trapping, hunting, fishing, timber and trespass laws, regulations and policies; be a law enforcement agent; know first aid; read and write letters, memos, reports and diaries; draw maps; administer timber sales; inventory wild animals and government property; install and repair tree and pole phone lines; mediate problems between the public and the Government; plant trees and dozens of other tasks.

Many men failed but those who made it usually stuck it out in spite of the low wages. The survival of the early Forest Service was their legacy. The ranger on the ground was the backbone of the outfit. They set the tone in public relations and finally acceptance of this new Government agency.

* * *

Buck continued as ranger on the Lakefield District for several years. In the mid-nineteen twenties he was offered the forest supervisor job at one of the largest and most complex workload national forests in the West. At the urging of many friends he took it. Of course, to him the only friend that really counted was Eddy. Typically, she said whatever Buck wanted to do was all right with her.

As supervisor, he was the boss of many professional rangers. He was a legend deserving total respect. No one ever mentioned the fact that he didn't have a college degree. It would have been a hollow statement. He knew more than any of the graduates on almost any aspect of the Service. In turn, he never mentioned how often or how badly many of them messed up. It was part of the learning process. He was asked to go to the Chief's office but turned it down with thanks. He still wanted to work near the field.

Buck retired during 1940, at the age of 64. Many times afterwards supervisor's would seek out his opinions and council. He usually turned them down, since he thought they should gain experience on their own, even if they fell flat on their face at times. The only time Buck would

Epilogue 233

talk or reminisce about his experiences was when about how they did something in the old days or if a question occurred relating to a historical fact. He still loved history and continued to read and study it all his life.

* * *

At Buck's retirement party in the Harmony Community Hall, a couple hundred of his old friends and colleagues attended. They were there to honor a man who had made their jobs easier, a man who they could look up to and a man whose decisions they remembered and benefited by during difficult times. Also in attendance were ranchers, loggers, miners and non-Forest Service admirers.

Buck wished Leroy Taylor and Lucy Neville could have been there. Both had passed away several years earlier. Lucy's funeral was the largest in the history of Harmony and Leroy's was the largest for Shadowcreek. They were special friends to Buck. He had attended both services.

His first two bosses, Kent Bolton and Garth Kimball, were present. The program lasted almost three hours. It seemed everyone had a story or two about Buck. Some stories were known but others were heard for the first time. There was no need to embellish the facts on any tale. The truth was enough.

Kent claimed he was the first person in the Forest Service who met Buck. He told the story about his fight in the bar and on the street and how Buck had rescued him. Garth mentioned how he was kidnapped for almost two days to put a roof on Buck's palace in the woods. Luke told the story for the first time about Buck's rescue of his neighbor, Mrs. Southcott, from the Crescent River. Alva described about growing up on a farm in Nebraska and about his brother capturing the two men who robbed him on the train to Shadowcreek.

There was an official envelope from the Chief's office in Washington, DC. Buck opened it and burst out laughing. It was a framed deed to Buck Stonewall, the current and only owner of an outhouse built in 1905, located at Lakefield Ranger Station site. It stated that Buck could either leave the remains alone or pick them up and use them wherever he wanted. It was assumed there was no other part of the surroundings the owner desired. From Kent's smile, Buck knew his old boss was responsible for the deed. There were other gifts from those in attendance and those in other offices around the country.

Edna spoke. It was a moving tribute to the man she loved. Her stories were of the domestic type. The times she had with Buck in the forest were personal and she wasn't about to share.

Finally it was Buck's turn. Looking around the room, his eyes filled with tears, he spoke. "You folks have given me the highest compliment a person can receive, by being here tonight. Yes, we had fun reliving those stories. Most of them were part of doing the job. Some of them even I had forgotten. I count each and every one of you as a friend and if ever our paths should cross after tonight, then I will be grateful each time. I loved my career in the Forest Service. The outfit made me settle down and realize there was an organization that I wanted to be part of. I didn't consider it a job. It was work but it wasn't a job. I enjoyed waking up and wondering what each day would bring. It was never the same."

Buck went around the room and said something about many in attendance. He thanked them all and ended with special recognition of Kent, Garth, Luke, Marshall, Tim and Alva.

Searching for the right words, Buck continued. "I also want to mention the best friend I ever had. He couldn't be here tonight, since he passed away several years ago. When that happened I cried for the third time in my life. The first time was when my older brother was killed. The second was after I buried my folks. And the third was when my faithful companion of many years went to horse heaven. Most of you remember Titus. He saved my life several times. He was a one of a kind horse for the ages and is buried in a marked grave at Singing Ranch."

He turned to Edna and their son, Thomas and daughter, Mary. "My son and daughter were Forest Service brats. They soon understood the demands of time and separation the work required but they adjusted well and to their credit and their mother's credit, they turned out well. Although Thomas didn't follow in the footsteps of his old dad, he is the foreman at the Reliance Lumber Company. Mary is married to John Watson, an influential local stockman most of you know. He couldn't be here tonight but their union has produced two grandchildren we love to spoil. I am proud to be the father of two such offspring.

"My biggest thanks goes to my dear, lovely wife, Eddy. None of this would have happened without Eddy by my side and that includes those first years when we were both single. She is the smartest person I have ever known. She has no fear. She is an outstanding mother and teacher. I would put her shooting prowess up against any of the men here tonight. Now I am going to brag a little. Many of you probably don't know she loves poetry and is a wonderful poet in her own right.

Epilogue

The Corona Publishing Company of Baltimore, Maryland, is going to issue a book of her poems. So please indulge me for a moment while I read a short example.

The Price

Each leaf is tethered to its twig,
It jerks and strains against the bond.
An elfin gust and it is free,
And being free is cut from life.
The wind enjoys the fatal game.
And aids each leaf to make the arcs
The cusps and the parabolas
Which mark its graceful path to earth.
Still it prances with the sweet breeze
And joins a group of dying leaves
The smell of truce is in the land.

I've talked long enough and you wonderful people are undoubtedly tired. Again thank you from the bottom of my heart. It has been a challenging and rewarding career. We should always be proud of the Forest Service and for what it stands. It will change with time but the basics will be there. Good night and God bless you."

There was a full two minutes standing ovation for the beloved ranger-supervisor.

* * *

As time passed, Buck and Edna continued to be active. They hiked, traveled, were loving grandparents and never lost their affection for the forest and its creatures.

One time Buck was asked to head a task force to report on the ecological impact of grazing, timber and watershed projects on the forests. It was a brilliant study and became a reference used for many years by the Service.

The Stonewalls moved back to Singing Ranch along the Crescent River. There they raised horses and made improvements on the barn, fences, landscaping and house. It was a showpiece of what could be termed a gentleman's farm.

Edna did volunteer work at the schools and church. She and Alice Picton continued their correspondence and friendship. Buck was asked to run for various offices in government but declined them all. Instead, he wrote a weekly article in the local newspaper named "The Rambling

Ranger." It was distributed to more than 20 other papers. One of his most requested pieces was in answer to a reader's question about what he enjoyed most being a ranger.

Buck wrote, "Other than the friendship of my fellow Forest Service brothers and sisters, the things that I enjoyed most of being a ranger took place on the Lakefield District. I rode thousands and thousands of miles along the creeks with their tumbling white water and deep pools murmuring to themselves and chattering on their long, lonely way to the sea. I rode over the mountains and saw the forest with the gigantic trees, soft mossy glens, pliable carpets of needles, little trees growing up trying to be big ones, struggling for 50, 60, 70 years to get their heads up in the light. Then, along the hilltops with the rock gardens of wildflowers arranged so beautifully. Overhead the blue, blue sky in the distance, ridge after ridge of seemingly endless mountains. Then down through the glaciated rocks carved out thousands of years ago, across the meadows into the lakes. There were hundreds and hundreds of camp fires, with wonderful companions; the talks—or just a dreamy gaze into the embers—and then to bed, under the stars. Thousands and thousands of bright dots on black velvet. And later, often the moon would come out, turning the whole mountainside into a vast fairyland, where the tinkling of the animal bells in the meadows was a musical accompaniment. Soon light would hit the first peak, joined by the drowsy chattering of birds in the trees. Often, as the sunbeams stole down through the trees, a squirrel would start scolding when he found his privacy was invaded. Then, as the light finally came through the forest, I'd know it was time to get up and go out for another glorious day.

"I knew it was fitting to leave my beloved district and move on, when the mountains started to become steeper and the days began to get longer. I knew it was time to go and make room for younger men with more energy and new ideas. The years in the Service after Lakefield were challenging and enjoyable but nothing could take the place of the close association with others working on the beautiful and bountiful land.

"Those were magical times on the Maahcooatche National Forest."

Notes

1. Currently the word postmaster is used for both males and females. (Page 5)
2. An Indian word meaning deer watering place (Mä kô AT chēē). (Page 6)
3. Jonathan was General Jackson's middle name. He gained the sobriquet "Stonewall" from his stand at Bull Run. (Page 11)
4. In 1935, the word forester was changed to chief. (Page 21)
5. A cast-iron pan with a handle used in frying food. Originally, it had long legs and was used over coals on the hearth. (Page 38)
6. A loose outer garment; an article of dress intended to be wrapped or fitted loosely around a person. (Page 56)
7. The *Use Book* uses the word district instead of allotment. (Page 57)
8. Cotton or linen cloth, printed with flowers and other devices in a number of different colors. (Page 63)
9. A cotton cloth with a printed pattern, usually glazed, used for curtains, etc. (Page 63)
10. The roots of the herb Arnica montana used in the form of a tincture and an embrocation for bruises, sprains, swelling, etc. (Page 63)
11. A single straight rod or staff, pointed and iron shod at the bottom for piercing the ground and having a socket joint at the top. It is used, instead of a tripod, for supporting a compass. (Page 66)
12. A steel instrument for paring the hoofs of horses. (Page 72)

13. In 1930, the word district was changed to region. (Page 91)

14. A puffy sleeve from the shoulder to the elbow having the general shape or outline of a leg of mutton. (Page 123)

15. A trimming of a piece of lace, chiffon or the like, usually ruffled or pleated and worn by women down the front of the dress. (Page 123)

16. A plaiting used for hats and bonnets made from leghorn straw grown in Tuscany, Italy, cut green and bleached. (Page 123)

17. A narrow ornamental fabric of silk, wool or cotton, often with a metallic wire, or sometimes a coarse cord, running through it, used as trimming for dresses, furniture, etc. (Page 144)

18. A gingham woven with colored warp and white filling yarns. Warp are the threads which are extended lengthwise in the loom and crossed by the woof or filling threads. (Page 144)

19. Artificial butter of oleomargarine, especially when made with the addition of butter. (Page 144)

20. Black became the standard color in 1913. (Page 152)

21. The mechanism, as a button or lever, controlling the discharge in a spark plug. (Page 152)

22. A muffin made with coarse flour. (Page 184)

23. Six tines one side. (Page 195)

24. A serenade played for a newly married couple, using pots and pans and horns, etc. (Page 214)

Names of People and Animals
(116)

Alice O'Neil .... Edna's friend
Alva Stonewall .... Buck's younger brother
Annie O. (Oakley) .... Markswoman

Bill Caltute .... Ranger
Bertha .... Co-owner of Mabel's Café
Black Pete .... Miner
Buck Stonewall .... Ranger
Bud Brown .... Character in report

Calvin Merton .... Clerk at Harmony Hotel
Calvin Nibbs .... Owner of Shadowcreek Mercantile
Cecilia (Risby) Hubbard .... Marshall Hubbard's wife
Charlie Shaw .... Dispatcher
Charlie Springer .... Ivy's son
Cliff Clayton .... Forest Supervisor
Collier Newton .... Reliance Lumber Company president
Collier Travers.................. General manager of TL&W Railroad

Doc Cleary .... Shadowcreek doctor
Dolly .... Harmony Boarding House manager

Ed Black .... Story character
Edna Lawrence .... Shadowcreek postmistress
Edward Pulaski .... Ranger
Eleanor Wells .... Philip Well's wife
Elisa Vaughan................... John Vaughan's wife

239

Elroy Taylor .... Ike's Livery Stable owner
Eugene Findley .... Preacher

Faye Wadsworth .... Pilot Hotel manager
Felicity Southcott .... Luke's neighbor
Fred Breen .... Letter writer

Garth Kimball .... Deputy forest supervisor
Gifford Pinchot .... First Forest Service forester
Green, Mr. .... Story character

Harris Dalton .... Rancher
Harvey Southcott .... Luke's neighbor
Henry Graves .... Second Forest Service forester
Henry Springer .... Ivy Stonewall's husband
Herman Woodford .... Outlaw
Hiram Lawrence .... Edna's first husband
Hugh Tanner .... Ranger

Ike Fraser .... Original owner Ike's Livery Stable
Ira Stonewall .... Buck's older brother
Ivy Stonewall .... Buck's sister

Jack Gavin, Mrs. .... Former Mrs. Rust
Jack Johnson .... Farmer
Jack Langtree .... Cook
Jake .... Cowboy
James Adair .... Special fiscal agent
Jeremy Conway .... Assistant ranger
Joe, Mrs. .... Wife of Old Joe
John Roster .... Owner Harmony Livery Stable
John Vaughan .... Assistant Ranger
John Watson .... Mary Stonewall's husband
Jonah Stratford .... Minister
Jonah Miller .... Miner
Jones, Mr. .... Story character
Jug Handley .... Stockman

Kent Bolton .... Forest supervisor

Names of People and Animals

Leland Peel	Injured hiker
Leo Pitman	Sheepman
Lloyd Knight	Letter writer
Lucas	Miner
Lucy Neville	Clerk
Luke Parley	Guard
Mabel	Co-owner Mabel's Café
Mae Dalton	Harris Dalton's wife
Mark Parley	Luke's son
Marquis of Queensberry	Patron of boxing
Marshall Hubbard	Assistant ranger
Mary Stonewall	Buck's and Edna's daughter
Matthew Parley	Luke's son
Mildred Denwick	Peter Blodgett's sister
Miles Rust	Timber operator
Nate Bennett	Ranger
Ned	Log Cabin bartender
Ned Buntline	Writer
Old Joe	Story character
Peter Blodgett	Rancher
Philip Wells	Pinchot's friend
Priscilla Hempstead	Ralph Hempstead's wife
Ralph Hempstead	Ranger
Ryan	Construction foreman
Sanford Picton	Acting forest supervisor
Sarah Parley	Luke's wife
Sheriff Miller	Queen's County sheriff
Sherman Clapham	Homesteader
Sidney Porter	Miner
Silas Pegler	Temporary guard
Smith, Mr.	Engineer
Stuart Brewer	Deputy forest supervisor
Susan Tanner	Hugh Tanner's wife

T. Shoemaker .... Forest Service employee
Tate .... Trapper
Theodore Roosevelt .... President
Thomas Stonewall .... Buck's and Edna's son
Thomas Stonewall Jackson .... General
Tim Westgate .... Ranger
Tiny Williams .... Saloon owner
Thornton Munger .... Forest Service employee
Tom McCarthy .... Guard
Tucker Foss .... Railroad dispatcher

William Daughs .... Invented split tree insulator
William Howard Taft .... President
White, Mr. .... District fiscal agent
White, Mrs. .... Story character

Zeke .... Bartender

Animals:

Baron .... Luke Parley's horse
Bucky............................ Marshall Hubbard's horse
Buddy .... Stuart Brewer's horse

Fiddler .... Kent Bolton's horse

Gin Rummy .... Sanford Picton's horse

Leo............................... Edna Stonewall's horse
Lightning .... Alva Stonewall's horse
Lily .... Jonah Miller's mule

Molly .... Edna Stonewall's dog

Nellie .... Buck Stonewall's dog

Stub .... Buck Stonewall's mule

Titus .... Buck Stonewall's horse

Geographical Names
(63)

Barnesville Ranger District and Station
Blackberry Creek
Bristol (town)

Canyon Springs District and Station
Clear Creek
Crescent Ranch
Crescent River
Crescent River School
Cuddahy Meadows (Allotment) District

Dalton Guard Station
Dalton Ranch

Finny Creek
Fish Creek Hill
Fish Cut (town)
Florin Peak
Florin Peak Lookout
Fulton District and Station

Golden Restaurant and Saloon
Green Lake

Harmony (town)
Harmony Boarding House

Harmony County Hospital
Harmony Hotel
Harmony Livery Stable
Holcomb Valley (Allotment) District
Hondo Indian Reservation
Horse Ridge
Humbug

Ike's Livery Stable
Ingot (town)

Johnson County
Joshua Peak

Lakefield District and Station
Log Cabin Bar
Long Tom District and Station
Lookingglass Creek

Maahcooatche National Forest
Mabel's Café
Morgan's Meadow
Morgan's Meadow Campground
Music Creek

Nelson Camp (Allotment) District

Peacock Tavern
Pilot Hotel
Pishi Rock (Allotment) District
Poker Flat (Allotment) District
Pondosa Ridge

Queens County
Queens County Courthouse

Rabbit Creek
Rocky Mountains National Forest

Shadowcreek (town)
Shadow Creek

Shadowcreek Community Church
Shadowcreek Post Office
Singing Ranch
Split Rock Timber Sale
St. Maries (town)

Tacoma (town)
Tip Top Mountain
TL& W Railroad Depot
Twin Flats

Wallace (town)

Acknowledgments

Many friends and a lifetime of experiences contributed to this book. Rangers have been sharing stories around their campfires since the Forest Service was founded. For the one about the skunk bear I owe thanks to Bruce Barron; to Jay Cravens for the railroad car; to Al Crebbin for his retirement speech; to Joe Ely for his story about Old Joe, a bear and the drift fence; to Les Joslin for the mule mix-up; to James R. Pratley for the holy land scene and to Jack Spencer for the two competitive mules.

Then, there were those who provided technical advice to help maintain the "it could have happened" basis of the book. They and their areas of expertise are:

Harry Frey:	Mining, homesteading and Indian allotments
Bill Jones:	Timber
Mac McEwen:	Telephones
Jay Mika:	Engineering and construction
Hank Mostovoy:	Fire
Shirley Mostovoy:	Clothing
Jim Rock:	Archaeology, Indians, food and log cabins
Dave Stetzel:	Surveying
Harry Taylor:	Grazing and range

All the poems except Edna's are from the book edited by John D. Guthrie listed in the bibliography. The beetle poem was written by S. W. Allen.

There are others who also helped in many ways. Their names are listed in alphabetical order: Grover C. Blake, Russ Bower, Jack Godden, Bob Gray, Chuck Lundeen, C. C. McGuire, Lee Morford, Mary Ellen Rock, Ralph S. Space, Richard Spray, Fred Wehmeyer and W. G. Weigle.

All the letters, reports and diaries were written by real people living during the book's time frame.

Thanks to Scollay Parker for his review of the text to ferret out errors and impossibilities.

Also thanks to Patricia O'Day who drew the illustrations and made several valuable recommendations.

A special thanks goes to my brother Dan Davies, a former newspaper editor, who read every word of the manuscript and made invaluable corrections and suggestions.

Finally, special kudos go to my partner Florice (Flo) Frank, another Forest Service retiree, who in the past has co-authored 13 books of travel and history with more than half relating to the Forest Service. She placed the entire manuscript on a computer and made her usual excellent comments.

Bibliography

Barron, Bruce, *Fabulous Memories of a Truly Adventurous Life*. Ox-Shoe Ranch Publications, Manton, CA, 2001.

Bergoffen, William W., *100 Years of Federal Forestry,* Forest Service, USDA, 1976.

Cravens, Jay H., *A Well Worn Path,* University Editions, Inc., Huntington, WV, 1994.

Davies, Gilbert W. & Frank, Florice M., *Stories of the Klamath National Forest-1905-1955*, HiStory ink Books, Hat Creek, CA, 1992.

Davies, Gilbert W. & Frank, Florice M., *Memories From the Land of Siskiyou,* HiStory ink Books, Hat Creek, CA. 1993.

Davies, Gilbert W. & Frank, Florice M., *Memorable Forest Fires,* HiStory ink Books, Hat Creek, CA, 1995.

Davies, Gilbert W. & Frank, Florice M., *Forest Service Humor,* HiStory ink Books, Hat Creek, CA, 1996.

Davies, Gilbert W. & Frank, Florice M., *Forest Service Memories*, HiStory ink Books, Hat Creek, CA, 1997.

Davies, Gilbert W. & Frank, Florice M., *Forest Service Animal Tales,* HiStory ink Books, Hat Creek, CA, 1998.

Frome, Michael, *Whose Woods These Are,* Doubleday & Company, Inc., Garden City, NY, 1962.

Gildart, Robert C., *Montana's Early-Day Rangers*, Montana Magazine, Inc., Helena, MT, 1985.

Gray, Bob, *Forests, Fires and Wild Things,* Naturegraph Publishers, Happy Camp, CA, 1985.

Gray, Gary Craven, *Radio for the Fireline*, Forest Service, USDA, 1982.

Guthrie, John D., *The Forest Ranger and Other Verse*, The Gorham Press, Boston, MA, 1919.

Joslin, Les, *Uncle Sam's Cabins*, Wilderness Associates, Bend, OR, 1995.

Joslin, Les, *Walt Perry, An Early-Day Forest Ranger in New Mexico and Oregon*, Wilderness Associates, Bend, OR, 1999.

Koch, Elers, *Forty Years a Forester*, Mountain Press Publishing Co., Missoula, MT, 1998.

McCulloch, Walter F., *Woods Words*, Oregon Historical Society and the Champoeg Press, 1958.

Morford, Lee, *Wildland Fires*, 1984.

Newland, James D., *Historic Resources Survey & Evaluation Report-Cleveland N.F.*, 1995.

Norris, Norman L., *The Magic of My Mountains*, Tule River Country Press, Springville, CA, 1998.

Peterson, C. O., *Those Other Years*, Professional Impressions, Darby, MT, 1992.

Pinchot, Gifford, *Breaking New Ground*, Harcourt, Brace, and Company, New York, NY, 1947.

Pratley, James R., *Forest Service Life*.

Robinson, Glen O. *The Forest Service*, Johns Hopkins University Press, Baltimore, MD, 1995.

Rock, Jim, *Log Cabins: Horizontal Log Construction*, Forest Service, USDA, 1979.

Rothman, Hal K., *I'll Never Fight Fire With My Bare Hands Again*, University Press of Kansas, Lawrence, KS, 1994.

Shaw, Charlie, *The Flathead Story*.

Space, Ralph S., *The Clearwater Story*, Forest Service, USDA, 1979.

Steen, Harold E., *The U. S. Forest Service A History*, University Washington Press, Seattle, WA, 1976.

Tixier, Stan, *Green Underwear*, Bonneville Books, Springville, UT, 2001.

Gilbert W. Davies, c 1968.

About the Author

Although this book is Gil Davies' first novel, it is the 24th published book he has either authored or edited, including six Forest Service books listed in the bibliography and two volumes of original cartoons. His 30 year career in the U. S. Forest Service included involvement in dozens of wildfires, history projects and development of handbooks, forms and directives for fires, contracting and property. The founding and development of the Klamath National Forest Interpretive Museum was one of his many endeavors.

A graduate of Willamette University (Salem, Oregon) with a degree in American history, he also attended graduate school at UCLA, majoring in meteorology. He worked in many professions including truck driver, airplane mechanic and hotel manager. Douglas Aircraft Company was his home for more than five years.

His interests include travel (46 states and more than 20 countries), books (more than 3,000 volumes in library), all the arts (plays piano and flute), and photography. He lives in the country along Hat Creek (between Mt. Shasta and Lassen Peak). He has three daughters, seven grandchildren and four great grandchildren.